P9-DWY-768

Plasma Dynamics

Plasma Dynamics

R. O. DENDY

CLARENDON PRESS · OXFORD
1990

'alton Street, Oxford OX2 6DP
' York Toronto
'cutta Madras Karachi
'ore Hong Kong Tokyo
Salaam Cape Town
Melbourne Auckland
and associated companies in
Berlin Ibadan

Oxford is a trade mark of Oxford University Press

Published in the United States
by Oxford University Press, New York

British Library Cataloguing in Publication Data
Dendy, R. O.
Plasma dynamics.
1. Plasmas
I. Title
530.44
ISBN 0-19-851991-5
ISBN 0-19-852041-7 pbk

Library of Congress Cataloging in Publication Data
Dendy, R. O.
Plasma dynamics/R. O. Dendy.
1. Plasma dynamics. I. Title.
QC718.5.D9D46 1990 530.4'4—dc20 89-48832
ISBN 0-19-851991-5
ISBN 0-19-852041-7 pbk

Typeset by The Universities Press (Belfast) Ltd
Printed in Great Britain by
Bookcraft (Bath) Ltd
Midsomer Norton, Avon.

Preface

The aim of this book is to explain the fundamental concepts of plasma physics. Anyone who has attended an undergraduate course in electricity and magnetism should be able to follow it: the most advanced background knowledge that is needed consists of Maxwell's equations, and these are reviewed in the Introduction. Throughout the book, emphasis is placed on the underlying physical principles, rather than on mathematical details. The plan of the book is as follows. In the Introduction, after a brief description of the nature of plasmas, we review Maxwell's equations and related topics. Then, in Chapter 1, we introduce the fundamental physical parameters that characterize a plasma: the plasma frequency, the Debye length, and Coulomb scattering parameters. Chapter 2 is concerned with the dynamics of plasma particles in a magnetic field. After defining the Larmor radius, cyclotron frequency, and guiding-centre position in a uniform field, we consider the effects of magnetic field inhomogeneity. The various classes of guiding-centre drift are discussed, and particle trapping is examined. In Chapter 3, we derive expressions for plasma properties by summing the behaviour of large numbers of individual particles. This approach, in which the plasma is treated as a conducting medium, leads naturally to the characterization of the different types of high-frequency wave. Chapter 4 introduces the subject of magnetohydrodynamics, in which the behaviour of plasma in a magnetic field is treated as a branch of fluid mechanics. We study a range of magnetohydrodynamic effects, such as flux freezing and Alfvén waves, which extend our understanding of plasma behaviour at low frequencies. The kinetic theory of plasmas is introduced in Chapter 5. This branch of the subject takes account of the spread of velocities among the plasma particles that arises when the temperature of the plasma is non-zero. Because detailed information on the velocity of each particle is not in practice available, plasma kinetic theory involves a statistical approach. We re-examine plasma waves from this viewpoint, and study the characteristic plasma phenomena of Landau damping and two-stream instability, together with topics which also have applications beyond plasma physics, such as negative energy waves and the Fokker–Planck equation. Finally, in Chapter 6, we study non-linear plasma physics. This yields further insight into the plasma phenomena that we studied earlier, and also introduces some of the general concepts of non-linear physics. These include solitons, modulational instability, and non-linear wave coupling. Thus, by the end of the book, the reader will have examined all the main methods of describing plasmas, and have become familiar with the fundamental concepts of plasma physics.

Oxford
June 1989

R.O.D.

Acknowledgements

It is a pleasure to thank my colleagues at Culham Laboratory and the Joint European Torus, particularly Dr C. N. Lashmore-Davies and Dr G. A. Cottrell, for their advice.

Contents

Introduction

The solid, liquid, and gaseous states of matter which occur at the surface of our planet are not typical of matter in the universe at large. Most of the visible matter in the universe exists as plasma, whereas lightning and the aurora are the only natural manifestations of the plasma state on Earth. In a plasma, the energy of the particles is so great that the electric forces which bind the atomic nucleus to its electrons are overcome. One might think of the resulting assembly of positively and negatively charged particles as a gas, but the waves that occur in it have much in common with waves in electrically conducting solids, and its bulk behavior is best described using terminology and equations developed for fluids: hence the fact that plasma is referred to as the fourth state of matter. For these reasons, when studying plasma physics, one often needs to rearrange and reapply concepts already acquired elsewhere, rather than deal with topics that are unfamiliar.

The sun, like most stars, is composed of plasma; in its core, the kinetic energy of the atomic nuclei, dissociated from their electrons, is so great that they can overcome their mutual electrical repulsion and fuse together, releasing energy. The attempt to reproduce on Earth the process of controlled thermonuclear fusion, coupled with the need to understand the behaviour of plasma in space, is at present the main impetus for research in plasma physics.

Before we embark on the main text, it is important to be aware of one of the distinctive features of plasma physics. A plasma consists of an assembly of charged particles, interacting with each other through the Coulomb force. The movement of each plasma particle is governed by the local electric field; at the same time, the particle is also a source of electric field. In order to see what happens in various physical situations, we shall need to obtain solutions which simultaneously satisfy the equation of motion and Maxwell's equations. This is known as the requirement of self-consistency, and we shall meet it again in many different contexts.

Let us now review the main concepts that we shall need, and refer to textbooks on electricity and magnetism for further details. First, there is Poisson's equation, which relates the electric field $\boldsymbol{E}$ to the electric charge density ρ_e:

$$\nabla \cdot \boldsymbol{E} = \rho_e / \varepsilon_o. \tag{I.1}$$

We use m.k.s. units; ε_0 is the dielectric permittivity of free space. The gradient operator is defined by

$$\nabla = \hat{\boldsymbol{e}}_x \frac{\partial}{\partial x} + \hat{\boldsymbol{e}}_y \frac{\partial}{\partial y} + \hat{\boldsymbol{e}}_z \frac{\partial}{\partial z},$$

where $\hat{\boldsymbol{e}}_x$, $\hat{\boldsymbol{e}}_y$, and $\hat{\boldsymbol{e}}_z$ are the usual unit basis vectors. Equation (I.1) is often used in combination with the divergence theorem, which applies to any vector field $\boldsymbol{A}(\boldsymbol{x})$ in a volume V enclosed by a surface S:

$$\int_V \nabla \cdot \boldsymbol{A}(\boldsymbol{x})\, \mathrm{d}^3\boldsymbol{x} = \int_S \boldsymbol{A}(\boldsymbol{x}) \cdot \mathrm{d}\boldsymbol{S}. \tag{I.2}$$

Here $\mathrm{d}^3\boldsymbol{x}$ is our notation for a volume element, and $\mathrm{d}\boldsymbol{S}$ denotes a vector surface element. Applying eqn (I.1), we can re-state Poisson's equation in the familiar form of Gauss' theorem. This relates the charge, Q, contained within a volume to the integral over its surface of the normal component of $\boldsymbol{E}$:

$$Q = \int_V \rho_\mathrm{e}\, \mathrm{d}^3\boldsymbol{x} = \varepsilon_0 \int_V \nabla \cdot \boldsymbol{E}\, \mathrm{d}^3\boldsymbol{x} = \varepsilon_0 \int_S \boldsymbol{E} \cdot \mathrm{d}\boldsymbol{S}. \tag{I.3}$$

Thus far, we have considered electrostatics. A magnetic field that changes in time can also act as a source for the electric field. This is described by Faraday's law of electromagnetic induction, which can be written

$$\nabla \times \boldsymbol{E} = -\frac{\partial \boldsymbol{B}}{\partial t}. \tag{I.4}$$

To see how eqn (I.4) arises, we need Stokes' theorem, which applies to any vector field $\boldsymbol{A}(\boldsymbol{x})$ on a surface S whose boundary is a contour C:

$$\int_S \{\nabla \times \boldsymbol{A}(\boldsymbol{x})\} \cdot \mathrm{d}\boldsymbol{S} = \oint_C \boldsymbol{A}(\boldsymbol{x}) \cdot \mathrm{d}\boldsymbol{l}. \tag{I.5}$$

Here $\mathrm{d}\boldsymbol{l}$ denotes a distance element that is directed tangentially to the contour C at the point $\boldsymbol{x}$. Applying eqn (I.5) to eqn (I.4), we obtain Faraday's relation between the rate of change of the magnetic flux $\Phi_m = \int_S \boldsymbol{B} \cdot \mathrm{d}\boldsymbol{S}$ through S, and the electromotive force that is induced around the boundary C:

$$\frac{\mathrm{d}\Phi_m}{\mathrm{d}t} = \int_S \frac{\partial \boldsymbol{B}}{\partial t} \cdot \mathrm{d}\boldsymbol{S} = \int_S (\nabla \times \boldsymbol{E}) \cdot \mathrm{d}\boldsymbol{S} = -\oint_C \boldsymbol{E} \cdot \mathrm{d}\boldsymbol{l}. \tag{I.6}$$

Note the similarity of the form in which we have written eqns (I.3) and (I.6). In both equations, the first and fourth expressions are quantities whose relations are relatively familiar. The middle expressions in eqns

(I.3) and (I.6) reformulate these quantities using vector analysis. The simpler of the remaining equations is

$$\nabla \cdot \boldsymbol{B} = 0. \tag{I.7}$$

Comparing eqns (I.1) and (I.7), we see that the sources of electric and magnetic fields differ fundamentally. Applying eqn (I.2), where S is necessarily a closed surface, to eqn (I.7), we obtain

$$\int_{S\ \text{closed}} \boldsymbol{B} \cdot \mathrm{d}\boldsymbol{S} = 0. \tag{I.8}$$

This is the familiar result that as much magnetic flux leaves any given volume as enters it. The last of Maxwell's equations relates the magnetic field to its two sources, the electric current density $\boldsymbol{J}$ and time-varying electric fields. For our purposes, we have

$$\nabla \times \boldsymbol{H} = \boldsymbol{J} + \varepsilon_0 \frac{\partial \boldsymbol{E}}{\partial t}. \tag{I.9}$$

Here we are using m.k.s. units, so that

$$\boldsymbol{B} = \mu_0 \boldsymbol{H}, \tag{I.10}$$

where μ_0 is the magnetic permeability of free space, related to ε_0 and to the velocity of light c through the identity

$$\mu_0 \varepsilon_0 = 1/c^2. \tag{I.11}$$

In eqns (I.1), (I.9), and (I.10), both the relative dielectric permittivity and the relative magnetic permeability have been set equal to unity. That is, the particles that compose the plasma, and give rise to the currents and charges, exist in empty space and not in some underlying medium.

If there are no time-varying electric fields, eqns (I.9), (I.10), and (I.5) give Ampère's law:

$$\oint_C \boldsymbol{B} \cdot \mathrm{d}\boldsymbol{l} = \mu_0 \int_S \boldsymbol{J} \cdot \mathrm{d}\boldsymbol{S}. \tag{I.12}$$

For example, it follows directly from eqn (I.12) that the magnetic field at a distance r from a wire carrying an electric current I has strength $B = \mu_o I/2\pi r$. We also note that for any vector field $\boldsymbol{A}$, the identity

$$\nabla \cdot (\nabla \times \boldsymbol{A}) = 0 \tag{I.13}$$

applies by the definition of the gradient operator ∇. Thus, any equation with $\nabla \times \boldsymbol{H}$ on the left-hand side must have a right-hand side whose divergence vanishes. Returning to eqn (I.9), and using eqn (I.1), we must have

$$\nabla \cdot \left(\boldsymbol{J} + \varepsilon_0 \frac{\partial \boldsymbol{E}}{\partial t} \right) = 0 = \nabla \cdot \boldsymbol{J} + \frac{\partial \rho_e}{\partial t}. \tag{I.14}$$

Integrating eqn (I.14) over a volume, and applying eqn (I.2) to the $\nabla \cdot \boldsymbol{J}$ integral, we obtain

$$\frac{\mathrm{d}Q}{\mathrm{d}t} = -\int_S \boldsymbol{J} \cdot \mathrm{d}\boldsymbol{S}. \tag{I.15}$$

This states the conservation of electric charge. The rate of change of charge within a volume is determined entirely by the flow of current through its surface.

There is only one remaining fundamental equation. It is the non-relativistic Lorentz force equation, which states that the force $\boldsymbol{F}$ on a particle of charge q in an electric field $\boldsymbol{E}$ and magnetic field $\boldsymbol{B}$ is

$$\boldsymbol{F} = q(\boldsymbol{E} + \boldsymbol{v} \times \boldsymbol{B}) \tag{I.16}$$

when the speed $v \ll c$. As we shall see, the entire subject of plasma physics is a demonstration of the range and power of this equation, together with Maxwell's equations.

1

Basic plasma characteristics

1.1 The electron plasma frequency

We start by considering an assembly that consists of an equal number of electrons with charge $-e$ and ions (atomic nuclei) with charge $+e$. Suppose that these particles are initially uniformly distributed, so that the plasma is initially electrically neutral everywhere. Furthermore, let us assume that there is no random thermal motion, so that the electrons and ions are initially motionless. This idealized assembly is referred to as a cold plasma. We now perturb this system by transferring a group of electrons from a given region of space, leaving a net positive charge behind, to a neighbouring region which acquires net negative charge; see Fig. 1.1. This local charge gives rise to an electric field $\boldsymbol{E}$. Since they are very much lighter than the ions, the electrons respond much more rapidly to the electric field, and the ion motion can be neglected. The electric field $\boldsymbol{E}$ acts on the electrons in such a way as initially to reduce the local charge non-neutrality that is its source. As the electric field accelerates the electrons, they acquire kinetic energy. When the electrons return to their initial positions, so that the plasma is again electrically neutral everywhere and $\boldsymbol{E}=0$, all the electrostatic potential energy associated with the initial perturbation has been converted to electron kinetic energy. This kinetic energy carries the electrons on, past their initial positions. The plasma again becomes non-neutral, and an electric field is set up which retards the electron motion. Eventually the electron velocity becomes zero; all the kinetic energy has been converted into the electrostatic energy associated with the charge non-neutrality that the movement of the electrons back past their equilibrium positions has created. The situation is now identical to that produced by the initial perturbation, except that the direction of $\boldsymbol{E}$ has reversed.

We have just outlined the first half cycle of an oscillatory process that could in principle continue indefinitely, as energy shifts from the electrostatic field to electron kinetic energy and back again, for ever. The frequency with which this oscillation occurs is known as the electron plasma frequency ω_{pe}. We can calculate ω_{pe} using the equation of motion for a single electron in the presence of an electric field

$$m\ddot{x} = -eE, \tag{1.1}$$

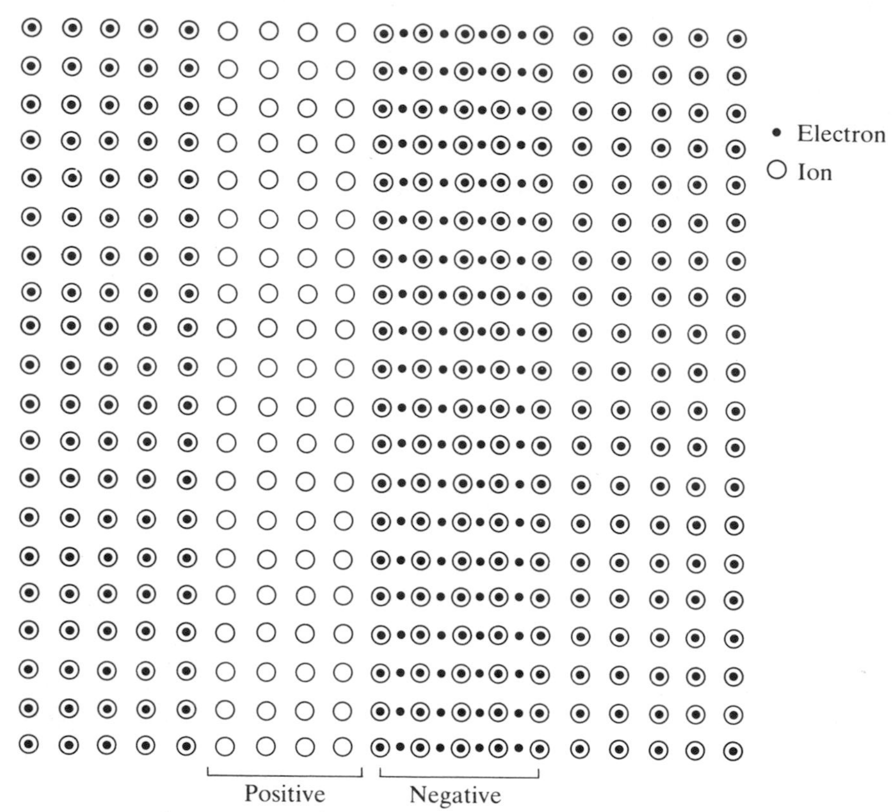

Fig. 1.1 Displacement of electrons in a cold plasma.

where x is the direction parallel to the electric field, and Gauss' Theorem

$$\int_S \boldsymbol{E} \cdot \mathrm{d}\boldsymbol{S} = Q/\varepsilon_0, \tag{1.2}$$

where Q is the charge contained within a closed surface S, as in eqn (I.3). In Fig. 1.2, the regions of positive and negative charge correspond to those in Fig. 1.1. Both figures are two-dimensional, as we assume that the configuration of the plasma particles does not depend on their position on the third spatial axis, which can therefore be suppressed. For the surface S shown in Fig. 1.2, we see that

$$\int_S \boldsymbol{E} \cdot \mathrm{d}\boldsymbol{S} = -LE. \tag{1.3}$$

Also, if n_0 is the equilibrium particle number density,

$$Q = -Lxn_0e \tag{1.4}$$

where x denotes the displacement of the electrons from their equilibrium positions as shown in Fig. 1.2. Substituting eqns (1.3) and (1.4) into eqn (1.2), we obtain $E = n_0ex/\varepsilon_0$, so that eqn (1.1) gives

$$\ddot{x} = -\omega_{\mathrm{pe}}^2 x, \tag{1.5}$$

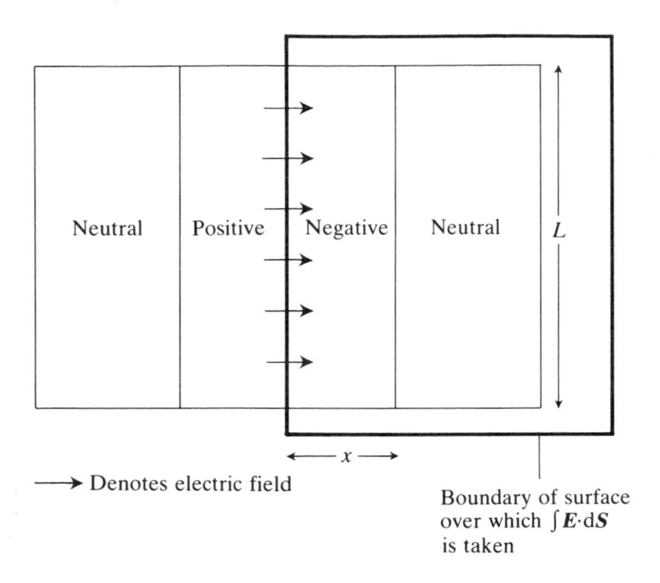

Fig. 1.2 Electric field in the plasma.

where

$$\omega_{pe} = (n_0 e^2/m\varepsilon_0)^{\frac{1}{2}} \tag{1.6}$$

is the plasma frequency. To obtain quantitative information, we make two further steps. First, we replace the universal constants e^2, m, and ε_0 by their numerical values. Secondly, we write n_0 as (n_0/characteristic density) $\times$ characteristic density. We choose the characteristic density to be that of a plasma in a present-day medium-size fusion experiment, $10^{19}\,\mathrm{m}^{-3}$. This enables us to write ω_{pe} given by eqn (1.6) in the form

$$\omega_{pe} \simeq 1.8 \times 10^{11}\left(\frac{n_0}{10^{19}\,\mathrm{m}^{-3}}\right)^{\frac{1}{2}} \mathrm{rad\,s^{-1}}. \tag{1.7}$$

Thus, when $n_0 = 10^{19}\,\mathrm{m}^{-3}$, we have $\omega_{pe} = 1.8 \times 10^{11}\,\mathrm{rad\,s^{-1}}$. In a diffuse plasma with $n_0 = 10^{11}\,\mathrm{m}^{-3}$, eqn (1.7) gives $\omega_{pe} = 1.8 \times 10^{7}\,\mathrm{rad\,s^{-1}}$.

This dependence of ω_{pe}^2 on the plasma and particle parameters is to be expected. The magnitude of the charge that is produced by an electron displacement x is, by eqn (1.4) and Fig. 1.2, proportional to the charge density $n_0 e$ which the removal of the electrons uncovers. This charge gives rise to the electric field in eqn (1.1), which is related to $\ddot{x}$ by the electron charge-to-mass ratio e/m. The fact that the electron mass m, though small, is non-zero, is fundamental to this discussion. If the electron had no mass, the electrostatic energy could not be transformed into electron kinetic energy. The fact that the electron mass is small accounts for the rapid response of the electrons to the electric field, and the correspondingly high value of ω_{pe}, which by eqn (1.7) is typically in the microwave range of frequencies for present-day fusion plasmas. Finally, we have used a cold plasma model, in which there is no random thermal motion, to derive eqns (1.5) and (1.6). These oscillations are therefore a completely ordered response of the plasma to electrostatic perturbations.

1.2 The Debye length

Let us return to the initial unperturbed cold plasma introduced in Section 1.1, and slowly insert into it a point positive charge, Q. The charge will attract a cloud of electrons, and repel the local ions, so that it is completely shielded from the rest of the plasma. Outside the cloud, there will be no electric field. Now raise the temperature of the electrons and ions from absolute zero to T, so that they acquire random thermal motion. Deep inside the cloud, the random thermal motion of the electrons will not be sufficient to enable them to escape from the vicinity of the point charge Q. At the edge of the cloud, most of the charge Q is screened by the inner electrons. In this outer region, the thermal energy of the electrons may exceed their electrostatic potential in the field of the largely screened charge Q—see Fig. 1.3. Such electrons are able to escape from the cloud; to the extent that this occurs, Q is no longer so effectively screened.

We can find a self-consistent solution for the electrostatic potential ϕ, which arises from the test charge Q, and the response of the plasma to the presence of Q. Using Boltzmann's equation, $n(E) = n_0 \exp(-E/k_B T)$ where k_B is Boltzmann's constant, we can describe how the potential ϕ that is produced by Q tends to drive away the ions and attract the electrons:

$$n_i = n_0 \exp(-e\phi/k_B T); \qquad n_e = n_0 \exp(e\phi/k_B T). \tag{1.8a,b}$$

The charge density ρ_e is given by

$$\rho_e = e(n_i - n_e) = -2n_0 e \sinh(e\phi/k_B T), \tag{1.9}$$

where we have used the identity $\sinh(x) = \frac{1}{2}(e^x - e^{-x})$. In eqn (1.9), we have treated ϕ as a source of charge separation. However, we can also express charge separation as a source of ϕ. This is a typical example of the self-consistency of plasma behaviour that was mentioned in the Introduction. Using Poisson's equation, eqn (I.1), and the fact that $\boldsymbol{E} = -\boldsymbol{\nabla}\phi$, we write

$$\nabla^2\phi = -\rho_e/\varepsilon_0. \tag{1.10}$$

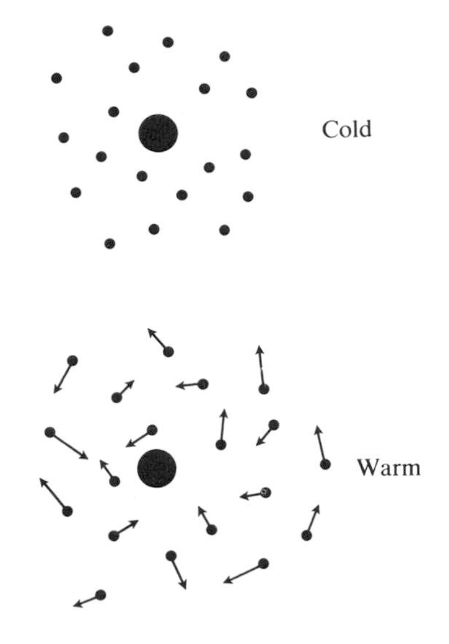

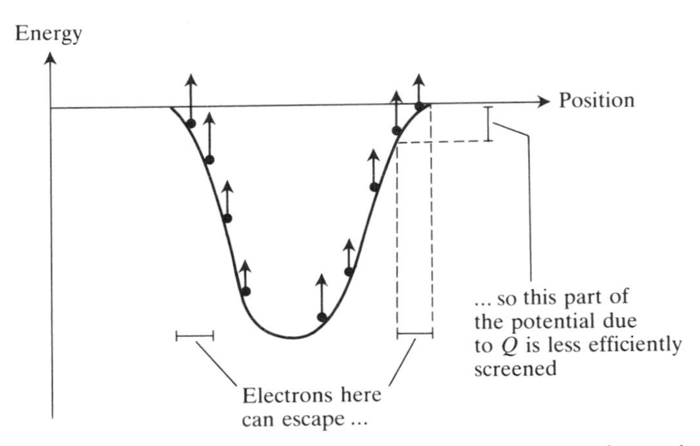

Fig. 1.3 Escape of electrons from screening cloud, due to thermal motion.

Substituting eqn (1.9) in eqn (1.10), we obtain the self-consistent result

$$\nabla^2\phi = \frac{2n_0 e}{\varepsilon_0}\sinh(e\phi/k_\mathrm{B}T). \tag{1.11}$$

Near the edge of the cloud, and beyond, the electrostatic energy $e\phi$ associated with Q is much less than the electron thermal energy $k_\mathrm{B}T$. We can then approximate $\sinh(e\phi/k_\mathrm{B}T) \simeq e\phi/k_\mathrm{B}T$, so that eqn (1.11) becomes

$$\nabla^2\phi = \frac{2}{\lambda_\mathrm{D}^2}\phi, \tag{1.12}$$

where we define the Debye length

$$\lambda_{\mathrm{D}} = \left(\frac{\varepsilon_0 k_{\mathrm{B}} T}{n_0 e^2}\right)^{\frac{1}{2}} = \left(\frac{k_{\mathrm{B}} T}{m}\right)^{\frac{1}{2}} \frac{1}{\omega_{\mathrm{pe}}}. \tag{1.13}$$

As with eqns (1.6) and (1.7), it is convenient to normalize the plasma parameters appearing in eqn (1.13). We again choose typical values appropriate to medium-size fusion experiments, with characteristic thermal energy 1 keV and number density $10^{19}\,\mathrm{m}^{-3}$. Thus eqn (1.13) can be written

$$\lambda_{\mathrm{D}} = 7.4 \times 10^{-5} \left\{ \left(\frac{k_{\mathrm{B}} T}{1\,\mathrm{keV}}\right) \Big/ \left(\frac{n_0}{10^{19}\,\mathrm{m}^{-3}}\right) \right\}^{\frac{1}{2}} \mathrm{m}. \tag{1.14}$$

It follows that, for example, the Debye length takes the value 7.4×10^{-5} m for a plasma at 1 keV with $n_0 = 10^{19}\,\mathrm{m}^{-3}$, and is 2.3×10^{-4} m for a plasma at 10 keV with the same density. The solution of eqn (1.12) is

$$\phi = \frac{Q}{r} \exp(-\sqrt{2}\, r/\lambda_{\mathrm{D}}). \tag{1.15}$$

By eqn (1.15), the Debye length λ_{D} is a measure of the range of the effect of the test charge Q. It follows from eqn (1.13) that this range is greater in a hot diffuse plasma than in a cool dense plasma. This is to be expected: if T is large, more electrons in the cloud at a given distance from Q will be able to escape, so that Q is less efficiently screened; if n_0 is small, electrons will have to be drawn from a larger volume in order to shield a given charge Q.

The Boltzmann distributions given by eqns (1.8a,b) can also be derived from arguments of particle dynamics, as follows. The force on a group of electrons is determined by the local pressure gradient and by the electric field. If the temperature is spatially uniform, we have*

$$nm\left(\frac{\partial v}{\partial t} + v\frac{\partial v}{\partial x}\right) = -k_{\mathrm{B}} T \frac{\partial n}{\partial x} + ne\frac{\partial \phi}{\partial x}. \tag{1.16}$$

Here v is the average velocity which describes the ordered bulk motion of the group of electrons. We seek a time-independent solution with $\partial/\partial t \to 0$, so that eqn (1.16) becomes

$$\frac{1}{n}\frac{\partial n}{\partial x} = \frac{-1}{k_{\mathrm{B}} T}\frac{\partial}{\partial x}\left(\frac{mv^2}{2} - e\phi\right). \tag{1.17}$$

It follows that n is proportional to $\exp[-\{(mv^2/2) - e\phi\}/k_{\mathrm{B}}T]$. In the absence of bulk motion, $v = 0$ and eqn (1.8b) follows.

The screening of electric charges that we discussed above has physical meaning only if a large number of particles are present in the screening cloud. If there is only a handful of screening particles, the charge Q

*The use of $\partial/\partial t + v\,\partial/\partial x$ as a total time derivative is discussed in detail on page 59.

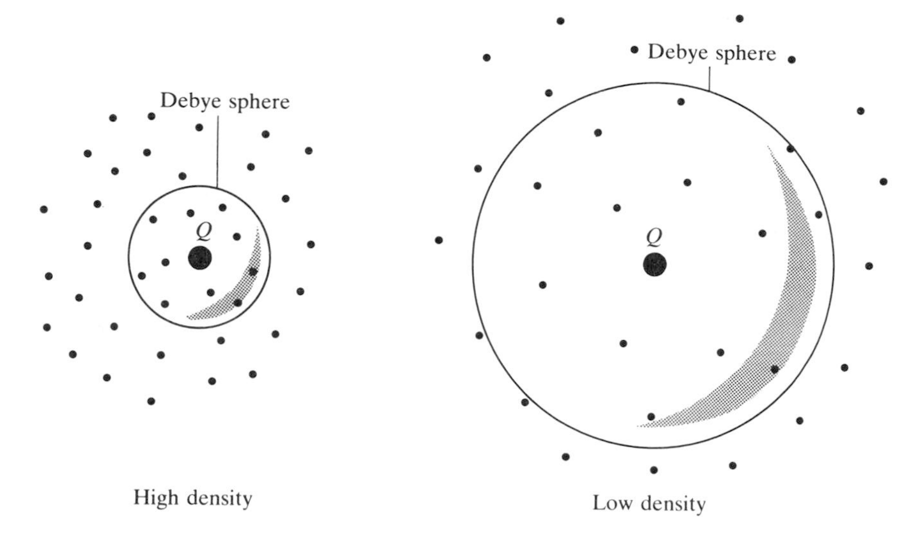

Fig. 1.4 Debye sphere.

remains unscreened in most directions at any given time. It is therefore useful to compute the number of particles N_D which lie within a Debye length of Q, and therefore participate in the screening. These particles are contained in a sphere which has radius λ_D and is centred on Q, known as the Debye sphere of Q (see Fig. 1.4). Its volume is $(4\pi/3)\lambda_D^3$, so that

$$N_D = \frac{4\pi}{3}\lambda_D^3 n_0 = 1.7 \times 10^7 \left(\frac{k_B T}{1\ \text{keV}}\right)^{\frac{3}{2}} \left(\frac{n_0}{10^{19}\ \text{m}^{-3}}\right)^{-\frac{1}{2}}. \tag{1.18}$$

We can now quantify the condition for effective screening to occur: it requires N_D to be large.

Finally, it has become clear from our discussion of ω_{pe} and λ_D that a plasma behaves in some respects as a classical dielectric medium. Overall, the plasma has no electric charge; individually, the plasma particles respond to applied fields and charges by moving to new positions. The effect of this movement is to set up new fields within the plasma that tend to counteract the applied field. We shall return to this basic dielectric aspect of plasma behaviour in Chapters 3 and 5.

1.3 Electrostatic plasma waves

We have so far discussed two aspects of the response of plasma electrons to perturbation. These are oscillation with characteristic frequency ω_{pe}, and charge screening with characteristic lengthscale λ_D. Both aspects play a role in electron wave motion in an unmagnetized plasma. Plasma wave motion is an example of the self-consistency mentioned in the Introduction, in the following sense. A group of electrons at a given point in the plasma will move in response to the wave field; by moving, they alter the local concentration of electric charge, thereby making their own contribution to the electric field. If the electrons are to participate in wave motion, they must oscillate coherently, in response to the local wave field. The constraint which this places on the range of wavelengths λ for which coherent wave motion is possible can be derived in two ways. First, note that the time which elapses before the oscillating electric field at a given point changes significantly is approximately ω_{pe}^{-1}. In this time, an electron moving at the average thermal velocity $v_T = (k_B T/m)^{\frac{1}{2}}$ travels a distance v_T/ω_{pe}, which by eqn (1.13) is just λ_D. If $\lambda_D \geqslant \lambda$, the electron samples the wave in all its phases, so that the forces acting on the electron cancel, and coherent oscillation is impossible. Only if

$$\lambda_D \ll \lambda \tag{1.19}$$

is the response of the electron dominated by the electric field in a particular restricted locality with a particular phase (see Fig. 1.5). Thus eqn (1.19) is the condition for a coherent response by the electrons. We can also obtain eqn (1.19) from screening arguments. For coherent oscillation, a given group of electrons must be influenced only by the local electric field of the wave. This requires more distant fields to be screened out, so that the electrons only respond to the field that is set up within a distance which is much less than the wavelength, which gives eqn (1.19) again. The condition eqn (1.19) becomes increasingly stringent as

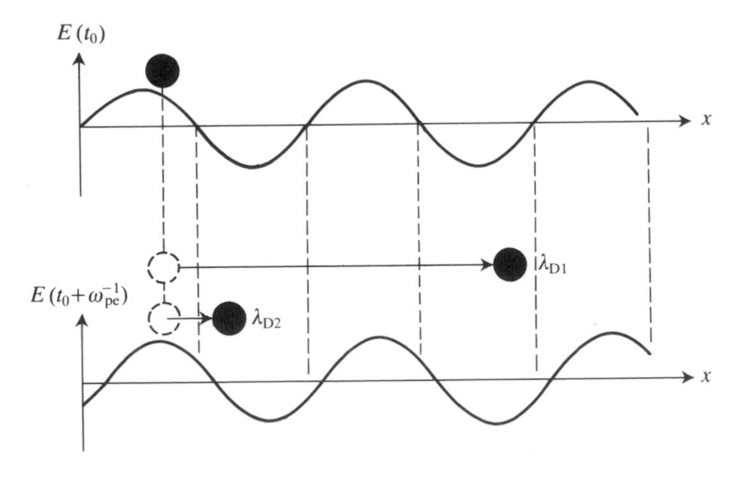

Fig. 1.5 Motion of an electron relative to a wave for two different values of the Debye length.

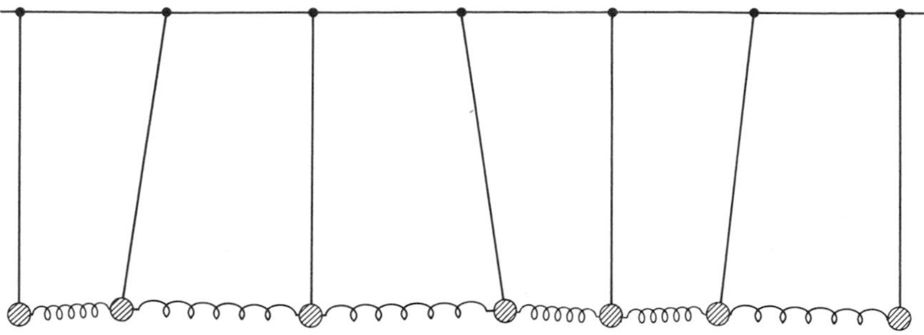

Fig. 1.6 Mechanical analogue of electron plasma waves.

the plasma grows hotter. This is because electric fields remain unscreened over a greater distance, since λ_D increases with temperature by eqn (1.13). Also, their increased thermal velocity enables the electrons to move further in the time ω_{pe}^{-1}.

The dynamics involved in electrostatic plasma waves can be modelled by a simple mechanical analogue, as follows. The dispersion relation* for a system of pendulums, which are all linked by identical springs to their neighbours, as in Fig. 1.6, is

$$\omega^2 = \frac{g}{l} + \left(\frac{4\pi^2 K a^2}{M}\right)(1/\lambda^2) \tag{1.20}$$

where g is the local acceleration due to gravity, K is the spring constant, and l, a, and M are respectively the length, separation, and mass of the pendulums. Equation (1.20) has the same form as the dispersion relation for electrostatic waves (also known as Langmuir waves) in unmagnetized plasma,

$$\omega^2 \simeq \omega_{pe}^2(1 + 3\lambda_D^2/\lambda^2), \qquad \lambda \gg \lambda_D \tag{1.21}$$

whose derivation will be postponed until we have discussed plasma kinetic theory in Chapter 5. In eqns (1.20) and (1.21), the role of g/l and ω_{pe}^2 shows that electrons are coupled by their charge to the electrostatic field set up by the rest of the plasma, just as the pendulums are individually coupled by their mass to the ambient gravitational field. Comparison of the $1/\lambda^2$ terms shows how λ_D appears in a pressure term arising from the finite temperature of the electrons, and independent of their charge; this is analogous to the influence on a given pendulum bob of all the other pendulum bobs, to which it is linked by the springs.

While it is usually the case that oscillating electric charges and currents give rise to electromagnetic radiation, the electron oscillations discussed here are exceptional for a reason which sheds further light on the concept of self-consistency. Recall that the magnetic source equation is given by

* A dispersion relation is a formula which gives the frequency of a wave in terms of its wavelength (or wavenumber).

eqn (I.9):

$$\nabla \times \boldsymbol{H} = \boldsymbol{J} + \varepsilon_0 \frac{\partial \boldsymbol{E}}{\partial t}. \tag{1.22}$$

The ions define a rest frame, and in this rest frame the electric current is $\boldsymbol{J} = -n_0 e \boldsymbol{v}$, where $\boldsymbol{v}$ is the velocity of the oscillating electrons. By eqn (1.1), $m\, \partial \boldsymbol{v}/\partial t = -e\boldsymbol{E}$ so that $\partial \boldsymbol{J}/\partial t = n_0 e^2 \boldsymbol{E}/m$. It follows that

$$\frac{\partial}{\partial t}\left(\boldsymbol{J} + \varepsilon_0 \frac{\partial \boldsymbol{E}}{\partial t}\right) = \varepsilon_0\left(\omega_{\mathrm{pe}}^2 \boldsymbol{E} + \frac{\partial^2 \boldsymbol{E}}{\partial t^2}\right) = 0, \tag{1.23}$$

where we have used eqn (1.6). There is therefore no source of magnetic field: the evolution in time of the electric field gives rise to a displacement current which exactly cancels the particle current associated with electron oscillation.

Finally, we note that the ions have so far been considered to be stationary. This is a good approximation, because the large mass of the ions in comparison to the electrons prevents the ions from responding rapidly to electric field oscillations. Of course, the movement of the ions cannot always be neglected. It is discussed, in different ways, in Chapters 3, 4, and 6.

1.4 Binary Coulomb collisions

So far in this discussion, the ions in the plasma have affected the dynamics of the electrons only by providing the overall electrical neutrality of the plasma. From time to time, however, a given electron must have a close encounter with an individual ion. This fact has not yet been included in our treatment of plasma dynamics, which has concentrated on collective behaviour rather than on binary interactions. We shall now examine the role of Coulomb collisions between single electrons and single ions, which can be treated as an instance of Rutherford scattering.

Figure 1.7 depicts the scattering of a charge q with mass m from a much heavier charge Q, which remains at rest. The charge q has initial velocity v_0 and impact parameter b, and is incident on a path which initially lies at an angle θ_0 to the z-axis, which is chosen to bisect symmetrically the completed particle orbit. The Coulomb force between the two charges has magnitude

$$F = qQ/4\pi\varepsilon_0 r^2, \tag{1.24}$$

and its component along the z-axis is $F_z = F\cos\theta$. The total momentum

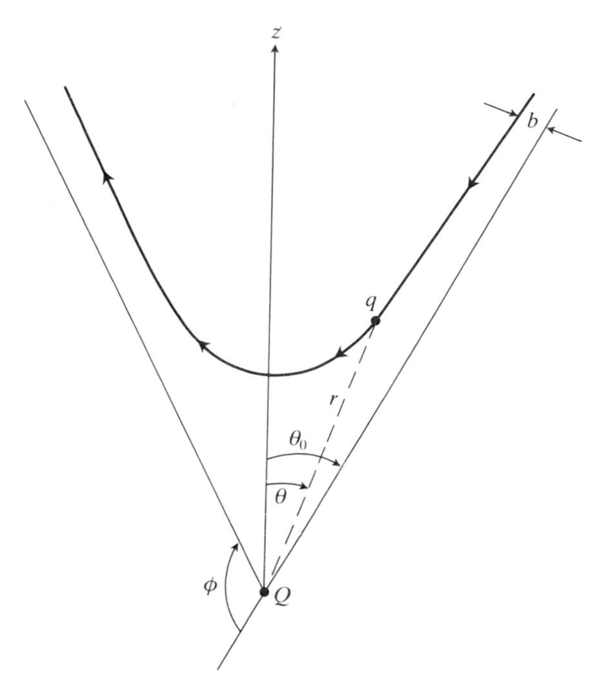

Fig. 1.7 Scattering of a charge q with impact parameter b relative to a much heavier charge Q.

imparted to q during scattering is

$$\begin{aligned}2mv_0\cos\theta_0 &= \int_{-\infty}^{\infty} F_z\,\mathrm{d}t = \int_{-\infty}^{\infty}\frac{1}{4\pi\varepsilon_0}\frac{qQ}{r^2}\cos\theta\,\mathrm{d}t\\ &= \frac{1}{4\pi\varepsilon_0}\int_{-\theta_0}^{\theta_0}\frac{qQ}{r^2\dot{\theta}}\cos\theta\,\mathrm{d}\theta\\ &= \frac{qQ\sin\theta_0}{2\pi\varepsilon_0 b v_0}.\end{aligned} \tag{1.25}$$

Here we have used the fact that the angular momentum $mr^2\dot{\theta}$ is constant in a central force field, and is always equal to its initial value mbv_0. It follows from Fig. 1.7 that the angle ϕ by which the charge q is deflected from its initial path is $\phi = \pi - 2\theta_0$. Using the identities $\sin(A+B) = \sin A\cos B + \cos A\sin B$ and $\cos(A+B) = \cos A\cos B - \sin A\sin B$, we have $\sin(\phi/2) = \cos\theta_0$ and $\cos(\phi/2) = \sin\theta_0$, so that

$$\cot(\phi/2) = \tan\theta_0. \tag{1.26}$$

Equation (1.25) yields an expression for $\tan\theta_0$, and substituting this into eqn (1.26) relates the scattering angle ϕ to the parameters of the Coulomb collision:

$$\cot(\phi/2) = 4\pi\varepsilon_0 mbv_0^2/qQ. \tag{1.27}$$

This is the standard Rutherford scattering formula.

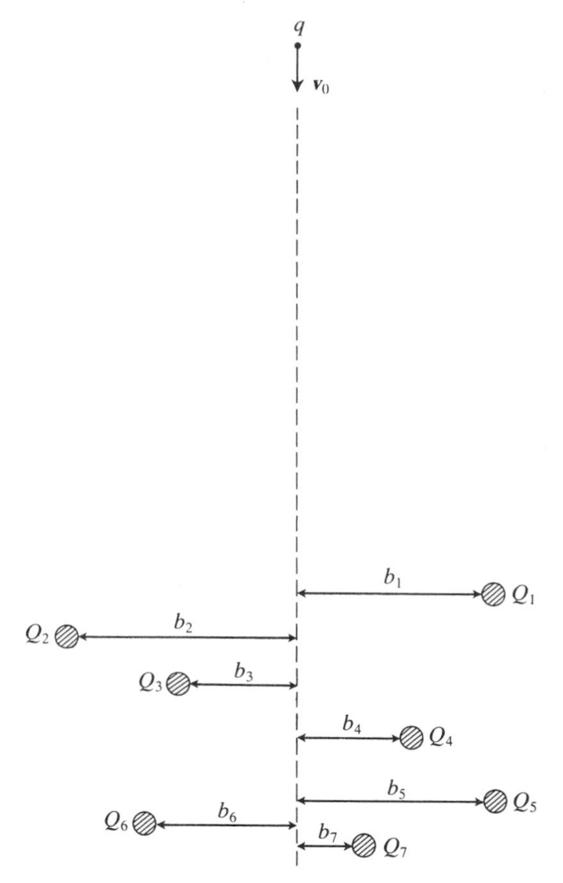

Fig. 1.8 Impact parameters of charge q with respect to a number of scattering centres.

Now $\cot(\phi/2) \rightarrow \infty$ as $\phi \rightarrow 0$. It follows that the impact parameter b associated with small-angle scattering rapidly becomes very large as increasingly small deflections ϕ are considered. There must therefore be a finite scattering angle for which the corresponding value of b is so large that q in fact has a smaller impact parameter with respect to some scattering centre other than Q. It follows that most charged particles in the plasma interact simultaneously and with comparable strength with a number of scattering centres, as in Fig. 1.8. This is a direct consequence of the long-range $1/r^2$ nature of the Coulomb force. We conclude that a simple investigation of binary Coulomb collisions indicates the desirability of a collective, self-consistent approach to plasma dynamics, of the kind illustrated in our treatment of ω_{pe} and λ_D.

In this chapter, we have introduced the fundamental parameters that characterize the collective, self-consistent behaviour of plasmas. First, we derived the natural frequency of oscillations of the number density of electrons, which is the electron plasma frequency ω_{pe}. This was calculated by self-consistent solution of Poisson's equation and of the electron equation of motion. Second, the Debye length λ_D was introduced. This is the lengthscale associated with the screening of electric charges by the electrons and ions. To calculate λ_D, we combined Poisson's equation and the Boltzmann distribution self-consistently. Next, we saw how ω_{pe} and λ_D enter the description of electrostatic plasma waves, and showed that, because of the cancellation of the electric current by the displacement current, the waves have no magnetic component. Finally, we examined Coulomb collisions between charged particles. The long-range nature of the Coulomb force causes the behaviour of each particle to be dominated by multiple Coulomb interactions with many other particles simultaneously. This fact led us back to our starting-point: the need for a collective, self-consistent description of plasmas.

Exercises

1.1. Complete the following table, which lists typical parameters for the electrons in four different types of plasma:

	n_0 ($\mathrm{m^{-3}}$)	T (K)	ω_{pe} ($\mathrm{rad\,s^{-1}}$)	λ_D (m)	N_D
Solar atmosphere	10^{18}	10^4			
Solar corona	10^{13}	10^6			
Ionosphere	10^{10}	10^3			
Tokamak	10^{19}	10^8			

1.2. Consider a spacecraft which passes through an electrically neutral hydrogen plasma, where the mean velocities of the electrons and ions are $\bar{v}_e$ and $\bar{v}_i$ respectively, with $\bar{v}_e \gg \bar{v}_i$, and the temperature of the electrons is T_e. Show that the spacecraft will acquire a negative electrical potential V, and that the maximum magnitude of V that should be allowed for is $\dfrac{k_B T_e}{e}\ln\left(\dfrac{\bar{v}_e}{\bar{v}_i}\right)$.

1.3. The set of coupled pendulums that is shown in Fig. 1.6 repays investigation, because of the analogy between the waves that occur in this system and electron plasma waves.

(a) Denoting the displacement from equilibrium of the nth pendulum by x_n, obtain an expression for $\mathrm{d}^2x_n/\mathrm{d}t^2$ in terms of x_{n-1}, x_n, and x_{n+1}, assuming that the latter are all small compared to the pendulum length l.

(b) Now make the continuous approximation: that is, assume that all neighbouring bobs have similar motion. This means that we may write the displacement of the bobs as a continuous function of their equilibrium position. Denoting the equilibrium position of the nth bob by z, that of the $(n-1)$th bob by $z-a$, and so on, we write the displacements in terms of a continuous function x as follows:

$$x_n(t) = x(z, t)$$

$$x_{n-1}(t) = x(z - a, t)$$

$$x_{n+1}(t) = x(z + a, t).$$

Note that this procedure replaces the discrete index n by a continuous position variable. Obtain the wave equation for this system, and show that it is consistent with eqn (1.21).

(c) What happens when we take the first bob in the system, and make it oscillate with an angular frequency below $(g/l)^{\frac{1}{2}}$?

Solutions are on pages 147 *and* 148

2

Motion of an electron in a magnetic field

2.1 Electron motion in a constant uniform magnetic field

Every plasma has a natural tendency to disperse. Unless there is some restraining force, the energetic particles that compose the plasma will travel away from their initial positions at high velocity and the plasma will cease to exist. Furthermore, many plasmas have low densities, and collisions are therefore so infrequent that a particle travelling outwards is unlikely to be deflected by a collision. Nevertheless, both in the laboratory and in space, diffuse collisionless plasmas can be sustained. They are prevented from dispersing by magnetic fields, which act on the charged particles through the Lorentz force. The dynamics of charged particles in magnetic fields therefore form an important area of plasma physics, which we shall examine in this chapter. We start by considering the simplest possible example, which consists of an electron moving in a spatially uniform magnetic field $\boldsymbol{B}$ which does not vary in time. In the absence of an electric field, the Lorentz force equation (I.16) simplifies to

$$m\dot{\boldsymbol{v}} = -e\boldsymbol{v} \times \boldsymbol{B}. \tag{2.1}$$

This equation can be written more conveniently if we introduce the electron cyclotron rotation vector

$$\boldsymbol{\omega}_{\mathrm{ce}} = \frac{e\boldsymbol{B}}{m}. \tag{2.2}$$

Combining eqns (2.1) and (2.2), we have

$$\dot{\boldsymbol{v}} = \boldsymbol{\omega}_{\mathrm{ce}} \times \boldsymbol{v}. \tag{2.3}$$

The magnitude of $\boldsymbol{\omega}_{\mathrm{ce}}$ is the electron cyclotron frequency, whose role we shall examine below. It is convenient to normalize B with respect to its value in a medium-size fusion plasma, and write

$$\omega_{\mathrm{ce}} = \frac{eB}{m} \simeq 1.8 \times 10^{11}\left(\frac{B}{1\ \mathrm{Tesla}}\right) \mathrm{rad\ s^{-1}}. \tag{2.4}$$

Comparing eqns (1.7) and (2.4), we see that under typical fusion plasma conditions, the frequencies ω_{pe} and ω_{ce} differ only by a factor of order unity. This coincidence is responsible for a degree of complexity in the behaviour of the plasma which does not arise if ω_{pe} and ω_{ce} differ greatly.

If we choose the direction of $\boldsymbol{\omega}_{ce}$ as the z-axis, eqn (2.3) shows that the acceleration $\dot{\boldsymbol{v}}$ has no z-component, so that v_z is constant. It also shows that $\boldsymbol{v} \cdot \dot{\boldsymbol{v}} = 0$, so that v^2 is constant. Physically, the magnetic field has no effect on the component of particle motion parallel to its own direction and, since the magnetic field gives rise to a force which is always perpendicular to the direction of particle motion, the particle energy cannot change. The most useful constant for describing the perpendicular motion is

$$v_\perp = (v^2 - v_z^2)^{\frac{1}{2}} \tag{2.5}$$

which is the magnitude of the velocity perpendicular to the magnetic field, $\boldsymbol{v}_\perp = \boldsymbol{v} - v_z \hat{\boldsymbol{e}}_z$; see Fig. 2.1. By eqn (2.3),

$$\dot{v}_x = -\omega_{ce} v_y, \qquad \dot{v}_y = \omega_{ce} v_x. \tag{2.6a,b}$$

Differentiating eqns (2.6a,b) with respect to time,

$$\ddot{\boldsymbol{v}}_\perp = -\omega_{ce}^2 \boldsymbol{v}_\perp. \tag{2.7}$$

Thus the perpendicular motion is composed of two independent simple harmonic motions, whose relative phase follows from eqns (2.6a,b):

$$v_x = v_\perp \cos(\omega_{ce} t + \phi), \qquad v_y = -v_\perp \sin(\omega_{ce} t + \phi), \tag{2.8a,b}$$

where ϕ is an arbitrary constant. Integrating eqns (2.8a,b) with respect to time, the coordinates of the electron are given by

$$x = x_0 + \frac{v_\perp}{\omega_{ce}} \sin(\omega_{ce} t + \phi), \tag{2.9a}$$

$$y = y_0 + \frac{v_\perp}{\omega_{ce}} \cos(\omega_{ce} t + \phi), \tag{2.9b}$$

$$z = z_0 + v_z t, \tag{2.9c}$$

corresponding to the helical path shown in Fig. 2.1. Here, the constants of integration (x_0, y_0) are the coordinates of the centre of a circular motion, and eqn (2.9c) follows from the constancy of v_z. The amplitude $v_\perp/\omega_{ce}$ that occurs in eqns (2.9a,b) is known as the Larmor radius of the electron:

$$r_L = \frac{v_\perp}{\omega_{ce}} \simeq 7.6 \times 10^{-5} \left(\frac{v_\perp}{1.3 \times 10^7 \text{ m s}^{-1}} \right) \left(\frac{B}{1 \text{ Tesla}} \right)^{-1} \text{m.} \tag{2.10}$$

By eqns (2.9a) and (2.9b), r_L is the distance from the centre of the circular motion to the electron itself. The normalizing velocity in eqn (2.10) is that of an electron of energy 1 keV. The dependence of r_L on $v_\perp$

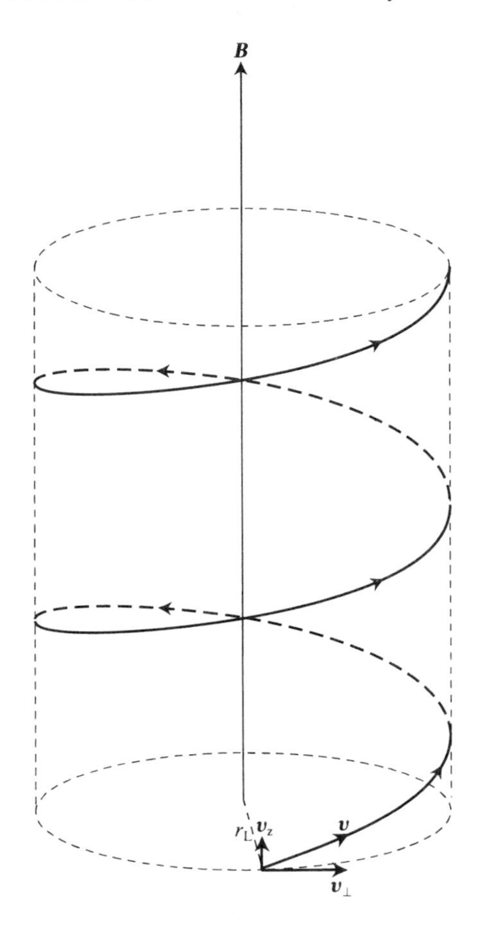

Fig. 2.1 Helical path of an electron in a uniform magnetic field.

and B follows from the fact that the magnetic field provides the centripetal force which holds an otherwise free electron in circular perpendicular motion. Thus r_L is the distance at which there is equilibrium:

$$\frac{mv_\perp^2}{r_L} = ev_\perp B \Rightarrow r_L = \frac{mv_\perp}{eB} = \frac{v_\perp}{\omega_{ce}}, \tag{2.11}$$

which is identical to eqn (2.10). In a weak magnetic field, the force causing the electron to deviate from straight-line motion is small, and the radius of curvature r_L of the electron path is correspondingly large. The $v_\perp$-dependence of eqn (2.10) follows from two considerations: the more energetic the electron, the more difficult it is for the magnetic field to deflect it; however, the magnetic force is itself proportional to the electron perpendicular velocity.

Finally, note that the circulating electron constitutes a current. If we project the motion of the electron onto a plane which is perpendicular to the magnetic field, a charge $-e$ passes a given point on its orbit every $2\pi r_L/v_\perp$ seconds, therefore the current $I = -ev_\perp/2\pi r_L$. The associated magnetic moment μ is the product of this current with the area enclosed by the orbit. Hence

$$\mu = \pi r_L^2 I = \frac{mv_\perp^2}{2B}. \tag{2.12}$$

The magnetic field to which μ gives rise opposes the applied field $\boldsymbol{B}$: this diamagnetic effect is, by eqn (2.12), proportional to the perpendicular energy.

2.2 Guiding centre drift due to non-magnetic forces

It follows from eqns (2.9a–c) that the path of an electron in a constant homogeneous magnetic field is helical. The helix is produced by uniform circular motion about a point that moves with constant velocity parallel to the magnetic field. It is useful to write the electron position as

$$\boldsymbol{x} = \boldsymbol{x}_c + \frac{\boldsymbol{v} \times \boldsymbol{\omega}_{ce}}{\omega_{ce}^2}, \tag{2.13}$$

which implicitly defines the guiding centre position $\boldsymbol{x}_c$; see Fig. 2.2. Differentiating eqn. (2.13) with respect to time, and using eqn (2.3) to eliminate $\dot{\boldsymbol{v}}$, we find $\dot{\boldsymbol{x}}_c = v_z \hat{\boldsymbol{e}}_z$ as required; note also that $|\boldsymbol{x} - \boldsymbol{x}_c| = r_L$. Now $\boldsymbol{x}_c$ is the mean position of the electron if the rapid rotation with frequency ω_{ce} is averaged out. This mean position often contains all the information required. By calculating $\boldsymbol{x}_c$, we can follow the path of the electron over a timescale which is long compared to ω_{ce}^{-1}. Consider, for example, an electron subjected to an impulsive collision, which by definition leaves its position unchanged but instantaneously changes $\boldsymbol{v}$ to $\boldsymbol{v} + \Delta\boldsymbol{v}$. Then

$$\begin{aligned} \boldsymbol{x} = \boldsymbol{x}_{c1} + \frac{\boldsymbol{v} \times \boldsymbol{\omega}_{ce}}{\omega_{ce}^2} &= \boldsymbol{x}_{c2} + \frac{(\boldsymbol{v} + \Delta\boldsymbol{v}) \times \boldsymbol{\omega}_{ce}}{\omega_{ce}^2} \\ \Delta\boldsymbol{x}_c = \boldsymbol{x}_{c2} - \boldsymbol{x}_{c1} &= \frac{\boldsymbol{\omega}_{ce} \times \Delta\boldsymbol{v}}{\omega_{ce}^2}. \end{aligned} \tag{2.14}$$

The instantaneous step $\Delta\boldsymbol{x}_c$ in guiding centre position perpendicular to the magnetic field is a measure of the effect of the instantaneous momentum transfer, see Fig. 2.3.

Consider now a continuous non-magnetic force $\boldsymbol{F}$ per unit mass. Instead of eqn (2.3), we have

$$\dot{\boldsymbol{v}} = \boldsymbol{\omega}_{ce} \times \boldsymbol{v} + \boldsymbol{F}. \tag{2.15}$$

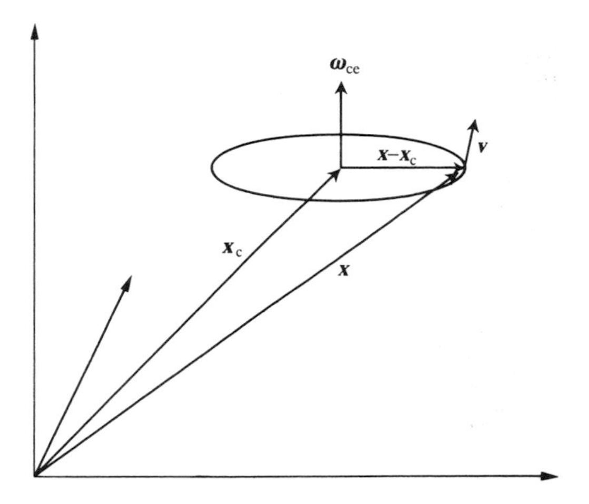

Fig. 2.2 Guiding centre position $\boldsymbol{x}_c$.

Differentiating eqn (2.13) with respect to time and using eqn (2.15),

$$\dot{\boldsymbol{x}}_c = v_z \hat{\boldsymbol{e}}_z + \frac{\boldsymbol{\omega}_{ce} \times \boldsymbol{F}}{\omega_{ce}^2}. \tag{2.16}$$

The second term on the right-hand side is the guiding centre drift velocity, $\boldsymbol{v}_d$. It is perpendicular both to the applied force and to the magnetic field. As an example, let us take an electric field, for which $\boldsymbol{F} = -e\boldsymbol{E}/m$. The

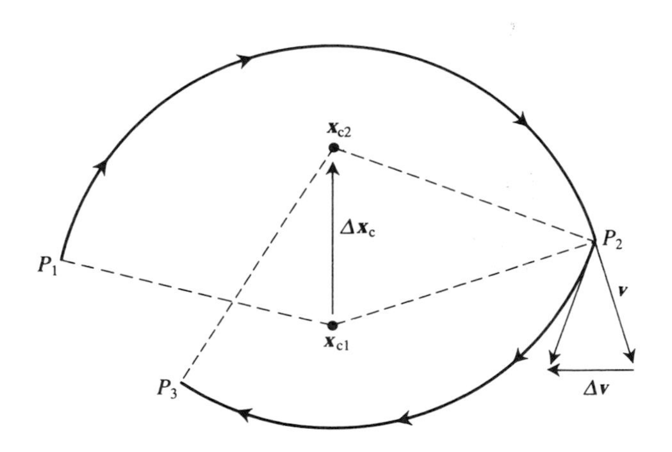

Fig. 2.3 Instantaneous guiding centre step $\Delta\boldsymbol{x}_c$ due to impulsive collision at P_2.

perpendicular drift of the guiding centre is

$$\boldsymbol{v}_\mathrm{d} = -\frac{e}{m}\frac{\boldsymbol{\omega}_\mathrm{ce}\times\boldsymbol{E}}{\omega_\mathrm{ce}^2} = \frac{\boldsymbol{E}\times\boldsymbol{B}}{B^2}\,. \tag{2.17}$$

This is a typical example of the three components of electron motion: rapid perpendicular rotation with frequency ω_ce at a radius r_L from $\boldsymbol{x}_\mathrm{c}$; parallel motion with constant velocity v_z; and a slow perpendicular drift $\boldsymbol{v}_\mathrm{d}$.

2.3 Guiding centre drift due to magnetic forces

In realistic plasma configurations, where the magnetic field is not uniform in space or time, the division of particle motion into three distinct components ceases to be exact. This division remains a useful approximation, however, so long as the variation of the magnetic field remains small on the scale of the cyclotron motion. That is,

$$(2\pi/\omega_\mathrm{ce})\frac{\partial B}{\partial t} \ll B, \tag{2.18a}$$

$$r_\mathrm{L}\,|(\boldsymbol{\nabla}B)_\perp| \ll B, \tag{2.18b}$$

$$v_z(2\pi/\omega_\mathrm{ce})\,|(\boldsymbol{\nabla}B)_\parallel| \ll B. \tag{2.18c}$$

We shall assume eqns (2.18a–c) to hold in the discussion which follows. This enables us to deal with non-uniformity in terms of a Taylor expansion about the particle orbit obtained for a uniform magnetic field.

Consider first a magnetic field whose strength increases in a direction perpendicular to the direction of the field itself. The field strength experienced by a particle changes periodically as it circulates with frequency ω_ce. The periodic shortening and lengthening of r_L causes the guiding centre to drift perpendicular to the magnetic field (see Fig. 2.4; for clarity, the gradient in magnetic field strength has been replaced by a discontinuity, so that this diagram actually violates eqn (2.18b)). The direction of this drift depends on the sense in which the particle is rotating, and is therefore opposite for electrons and ions. Let

$$\boldsymbol{B} = \left(B_0 + y\frac{\partial B_z}{\partial y}\right)\hat{\boldsymbol{e}}_z, \tag{2.19}$$

and define a scale length $l_c = \{(1/B_0)\,\partial B_z/\partial y\}^{-1}$. Using eqn (2.19) in eqn (2.1), and denoting the cyclotron frequency eB_0/m of the plasma at $y=0$ by ω_ce, the electron motion is governed by

$$\dot{v}_x = -\omega_\mathrm{ce}v_y - \frac{\omega_\mathrm{ce}}{l_\mathrm{c}}yv_y, \tag{2.20a}$$

$$\dot{v}_y = \omega_\mathrm{ce}v_x + \frac{\omega_\mathrm{ce}}{l_\mathrm{c}}yv_x. \tag{2.20b}$$

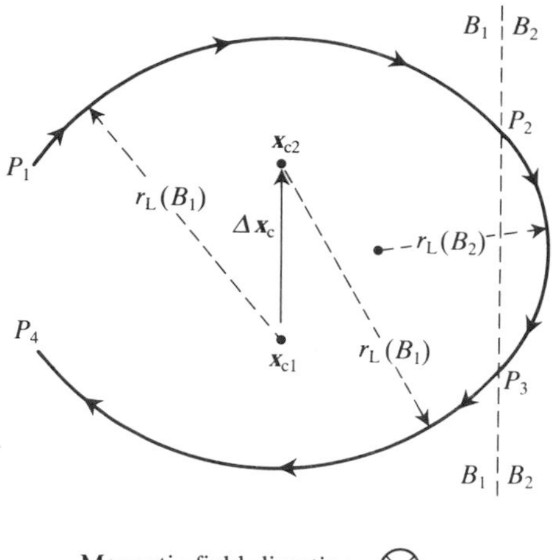

Fig. 2.4 Guiding centre step $\Delta \boldsymbol{x}_c$ due to non-uniformity of magnetic field.

In the uniform limit, $l_c \to \infty$ and eqns (2.20a,b) reduce to eqns (2.6a,b). We now use the uniform field orbit described by eqns (2.8a,b) and (2.9a,b) to calculate the effect of the additional terms in eqns (2.20a,b); setting $x_0 = y_0 = 0$ for convenience,

$$-\frac{\omega_{ce}}{l_c} y v_y = \frac{\omega_{ce} r_L v_\perp}{2 l_c} \sin 2(\omega_{ce} t + \phi), \tag{2.21a}$$

$$\frac{\omega_{ce}}{l_c} y v_x = \frac{\omega_{ce} r_L v_\perp}{2 l_c} (1 + \cos 2(\omega_{ce} t + \phi)). \tag{2.21b}$$

There is thus an extra non-oscillatory contribution to $\dot{v}_y$ only; all other terms arising from the spatial non-uniformity of the magnetic field strength average to zero over a single orbit. The average additional acceleration of the electron is $\boldsymbol{a} = (0, v_\perp^2/2l_c, 0)$. Substituting $\boldsymbol{a}$ for the force per unit mass $\boldsymbol{F}$ in eqn (2.16), it follows that the associated guiding centre drift is $\boldsymbol{v}_d = -(r_L v_\perp / 2 l_c)\hat{\boldsymbol{e}}_x$. Note that $\boldsymbol{v}_d$ is slower than $v_\perp$ by the factor $(2l_c/r_L)$, which by eqn (2.18b) is assumed large. The direction of $\boldsymbol{v}_d$ is perpendicular both to $\boldsymbol{B}_0$ and to ∇B. Since the choice of the y-axis for the direction of the perpendicular component of ∇B was arbitrary, we have in general

$$\boldsymbol{v}_d = \frac{r_L v_\perp}{2} \frac{\boldsymbol{B} \times \nabla B}{B^2}. \tag{2.22}$$

Now consider the case of a gradient in the magnetic field strength parallel to the direction of the magnetic field: $\partial B_z/\partial z \neq 0$. This variation in field strength cannot be the only deviation from exact uniformity. We recall from eqn (I.7) that $\nabla \cdot \boldsymbol{B} = 0$, so that in terms of cylindrical coordinates (R, θ, z) we have

$$\frac{1}{R}\frac{\partial}{\partial R}(RB_R) + \frac{\partial B_z}{\partial z} = 0 \Rightarrow B_R = -\tfrac{1}{2}R\left(\frac{\partial B_z}{\partial z}\right). \tag{2.23}$$

This small radial field is crucial to the particle dynamics of the system. It is perpendicular to the main component of the field, and to the orbit of the electrons in the B_z field, since $\boldsymbol{v}_\perp$ is in the $\hat{\boldsymbol{e}}_\theta$-direction. While B_R is zero on the axis of symmetry by eqn (2.23), an electron senses a finite value of B_R as it orbits at a distance $R = r_{\mathrm{L}}$ from the magnetic field line on which its guiding centre lies, see Fig. 2.5. It therefore experiences an additional acceleration

$$\boldsymbol{a} = \frac{-e}{m} v_\perp \hat{\boldsymbol{e}}_\theta \times \left(-\tfrac{1}{2} r_{\mathrm{L}} \frac{\partial B_z}{\partial z} \hat{\boldsymbol{e}}_R\right) = \frac{-v_\perp^2}{2 l_{\mathrm{c}}} \hat{\boldsymbol{e}}_z, \tag{2.24}$$

where we have defined a scale length $l_{\mathrm{c}} = \{(1/B_0)\,\partial B_z/\partial z\}^{-1}$ and used eqn (2.10). Note that the same basic parameters, $v_\perp^2$ and l_{c}, appear here and in the case of a perpendicular gradient. The acceleration is independent of charge: in eqn (2.24), the effects of $v_\perp \hat{\boldsymbol{e}}_\theta \to -v_\perp \hat{\boldsymbol{e}}_\theta$ and $-e \to e$ cancel. The effect of the acceleration is to reduce the parallel velocity of a particle as it moves into a region of higher field strength. Combining eqns (2.12) and (2.24), the force is

$$F_z = -\mu \frac{\partial B}{\partial z}. \tag{2.25}$$

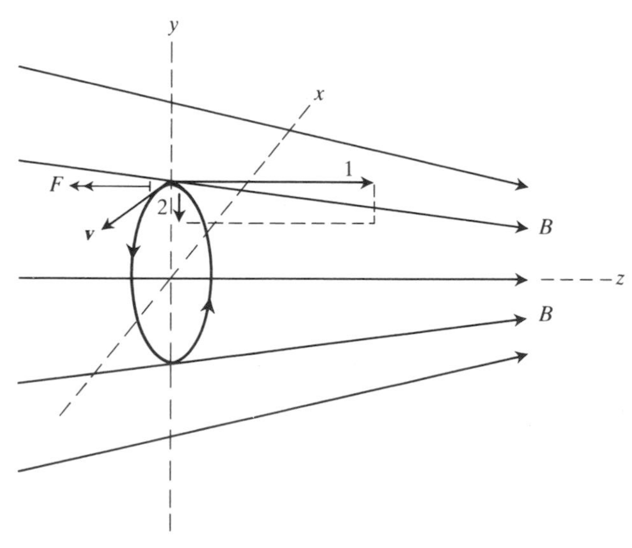

Fig. 2.5 Magnetic field sensed by orbiting electron when $\partial B_z/\partial z \neq 0$.

This force is purely magnetic in origin. Therefore, as pointed out in Section 2.1, it cannot change the total electron energy. When v_z decreases owing to F_z, $v_\perp$ must increase:

$$F_z v_z = -v_z \frac{\partial}{\partial z}(\tfrac{1}{2}mv_\perp^2), \tag{2.26}$$

where $v_z\, \partial/\partial z$ gives the rate of change with time following the guiding centre motion.

Now consider a magnetic field that varies with time. This necessarily induces an electric field in the way described by eqn (I.4):

$$\nabla \times \boldsymbol{E} = -\frac{\partial \boldsymbol{B}}{\partial t}. \tag{2.27}$$

Any parallel component $\boldsymbol{E}_z$ will accelerate particles along the magnetic field lines. In addition, if $\boldsymbol{E}$ includes a perpendicular component, the particle perpendicular energy may be changed as follows. In the time $T_0 = 2\pi/\omega_{\text{ce}}$ taken for one cyclotron gyration, the energy acquired by an electron is

$$\begin{aligned}
\Delta(\tfrac{1}{2}mv_\perp^2) &= -e \oint_{\text{orbit}} \boldsymbol{E} \cdot \mathrm{d}\boldsymbol{l} = -e \int_{\text{S}} \nabla \times \boldsymbol{E} \cdot \mathrm{d}\boldsymbol{S} \\
&= e \int \frac{\partial \boldsymbol{B}}{\partial t} \cdot \mathrm{d}\boldsymbol{S} = e\pi r_{\text{L}}^2 \frac{\partial B}{\partial t} \\
&= \mu T_0 \frac{\partial B}{\partial t},
\end{aligned}$$

where we have used eqns (2.12), (2.27), and Stokes' theorem. It follows that

$$\frac{\partial}{\partial t}(\tfrac{1}{2}mv_\perp^2) = \mu \frac{\partial B}{\partial t}. \tag{2.28}$$

We can now consider how magnetic fields, for which both $\partial B/\partial t$ and $\partial B/\partial z$ are non-zero, affect the perpendicular motion of the electron. Combining eqns (2.25), (2.26) and (2.28),

$$\frac{\mathrm{d}}{\mathrm{d}t}(\tfrac{1}{2}mv_\perp^2) = \mu \frac{\mathrm{d}B}{\mathrm{d}t}. \tag{2.29}$$

where $\mathrm{d}/\mathrm{d}t = \partial/\partial t + v_z\, \partial/\partial z$ is the total rate of change with time following the guiding centre. Since $mv_\perp^2/2 = \mu B$ by eqn (2.12), it follows from eqn (2.29) that μ, and hence $v_\perp^2/B$, is a constant of the particle motion; it is sometimes known as the first adiabatic invariant. The magnetic flux passing through the orbit of the electron is

$$\int \boldsymbol{B} \cdot \mathrm{d}\boldsymbol{S} = \pi r_{\text{L}}^2 B_z = \pi \left(\frac{m}{e}\right)^2 \frac{v_\perp^2}{B}, \tag{2.30}$$

which is proportional to μ and therefore constant. Some of the implications of this result will be examined in Section 2.5. Before we do so, there is one other aspect of guiding centre drift motion to be considered.

2.4 Curvature drift

The drift motions that we have described so far are of a particular kind. They originate in the fact that electrons have finite Larmor radius, and therefore sample magnetic field inhomogeneity in a small region with radius r_L around the guiding centre. The averaged effect of this sampling process has been expressed in terms of a drift of the guiding centre. We have carried out these calculations with respect to a frame, depicted in Fig. 2.2, which we have assumed is not itself subject to acceleration. That is, the frame is an inertial one, which means that particle accelerations are parallel to the applied force. Suppose now that the magnetic field lines are curved. The guiding centre motion will tend to follow the curved field line. Since we are concerned with the deviation of particle motion from magnetic field lines, it is useful to deal with the particle motion using a frame which is also following the curved field line. This enables us to leave implicit the fact that the field line is curved and to deal with the drifts that we have previously described in the usual way. We carry out this transformation of frames as an alternative to describing in mathematical detail the curved path of the magnetic field line. However, the effect of curvature cannot be entirely transformed away. The frame that we are now using is an accelerated one, and is subject to a centripetal force associated with the curvature of the path that it is following. This frame is therefore non-inertial, and the acceleration of particles measured in this frame is not parallel to the externally applied force $\boldsymbol{F}$. We can see this using Fig. 2.6: here O is the origin of an inertial frame, and $\boldsymbol{R}$ is the position of a test particle of mass m with respect to O; O′ is the origin of a non-inertial frame, which follows the curved path shown; $\boldsymbol{c}$ is the instantaneous position of O′ with respect to O, and $\boldsymbol{r} = \boldsymbol{R} - \boldsymbol{c}$ is the position of the test particle with respect to O′. Suppose that an externally applied force $\boldsymbol{F}$ is acting on the test particle. Then Newton's Second Law, applied in the inertial frame, gives

$$\boldsymbol{F} = m\ddot{\boldsymbol{R}}. \tag{2.31}$$

Now $\ddot{\boldsymbol{R}} = \ddot{\boldsymbol{r}} + \ddot{\boldsymbol{c}}$ by the definition of $\boldsymbol{R}$, $\boldsymbol{c}$, and $\boldsymbol{r}$. So by eqn (2.31), the mass acceleration $m\ddot{\boldsymbol{r}}$ of the test particle in the non-inertial frame is given by

$$m\ddot{\boldsymbol{r}} = \boldsymbol{F} - m\ddot{\boldsymbol{c}}. \tag{2.32}$$

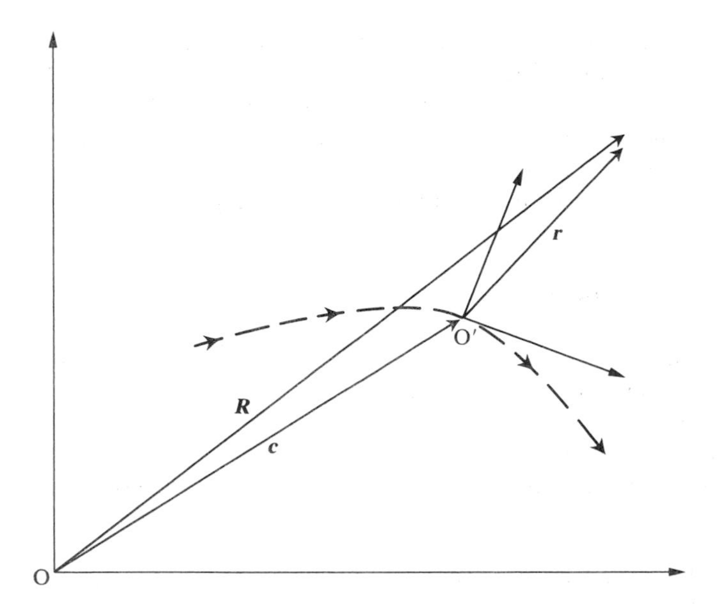

Fig. 2.6 Test particle position with respect to inertial and non-inertial frames.

Unlike the acceleration $\ddot{\boldsymbol{R}}$ in the inertial frame, the acceleration $\ddot{\boldsymbol{r}}$ in the non-inertial frame is not parallel to the externally applied force $\boldsymbol{F}$. Instead, it is parallel to the resultant of $\boldsymbol{F}$ and the force

$$\boldsymbol{F}_{\mathrm{c}} = -m\ddot{\boldsymbol{c}} \tag{2.33}$$

which depends on the non-inertial frame itself.

In our case, $\boldsymbol{F}_{\mathrm{c}}$ is the centripetal force; $\ddot{\boldsymbol{c}}$ is determined by the velocity $v_{\parallel}$ of the particle along the field line, which the non-inertial frame follows, and by the local radius of curvature R_{c} of the field line. Its magnitude is $\ddot{c} = v_{\parallel}^2/R_{\mathrm{c}}$. The centripetal force $\boldsymbol{F}_{\mathrm{c}}$ is a continuous non-magnetic force, and hence the drift to which it gives rise can be calculated using eqn (2.16). Suppose, for example, that the field lines are predominantly in the z-direction, but are slightly curved by a component in the x-direction. Then $R_{\mathrm{c}}^{-1} = (\partial B_x/\partial z)/B$, so that $\ddot{\boldsymbol{c}} = (v_{\parallel}^2(\partial B_x/\partial z)/B)\hat{\boldsymbol{e}}_x$. Combining eqns (2.16) and (2.33), this gives the curvature drift of the guiding centre:

$$\boldsymbol{v}_{\mathrm{d}} = \frac{\boldsymbol{\omega}_{\mathrm{ce}} \times \boldsymbol{F}_{\mathrm{c}}}{\omega_{\mathrm{ce}}^2} = -\frac{mv_{\parallel}^2}{B}\left(\frac{\partial B_x}{\partial z}\right)\frac{\boldsymbol{\omega}_{\mathrm{ce}} \times \hat{\boldsymbol{e}}_x}{\omega_{\mathrm{ce}}^2}. \tag{2.34}$$

Since $\boldsymbol{\omega}_{\mathrm{ce}} = (eB/m)\hat{\boldsymbol{e}}_z$, we see that the direction of the curvature drift $\boldsymbol{v}_{\mathrm{d}}$ is perpendicular to the plane defined by the local magnetic field direction and its curvature vector. The motion of the guiding centre in the inertial frame is the vector sum of the motion of the non-inertial frame, and the motion of the guiding centre in that frame. We conclude that a curvature drift given by eqn (2.34) will be observed whenever magnetic field lines are curved. We note also from eqn (2.34) that the curvature drift is a perpendicular quantity whose magnitude depends partly on $v_{\parallel}$. Thus magnetic field line curvature has the effect of coupling the perpendicular and parallel components of guiding centre motion.

2.5 Particle trapping

Recall from Section 2.3 that in a magnetic field that varies slowly in space and time, the magnetic moment $\mu = (m/2)(v_{\perp}^2/B)$ is a conserved quantity of the particle motion. Consider a static inhomogeneous magnetic field, with no applied electric field. Then the particle kinetic energy $K = (m/2)(v_{\perp}^2 + v_z^2)$ is also constant. Let us denote by subscript zero the values of particle and field parameters at the initial position, which for convenience we shall locate at the point where the magnetic field strength is weakest. We have in general, by conservation of μ and K,

$$v_{\perp}^2 = (B/B_0)v_{\perp 0}^2, \tag{2.35}$$

$$v_z^2 = v_0^2 - v_{\perp}^2 = v_0^2\left(1 - \frac{B}{B_0}\frac{v_{\perp 0}^2}{v_0^2}\right). \tag{2.36}$$

The value of $(v_{\perp 0}^2/v_0^2) < 1$ is fixed for a given particle. We know from eqn (2.25) that the value of v_z^2 diminishes as the particle moves into a region of stronger magnetic field, and we have seen how this arises from the condition $\boldsymbol{\nabla} \cdot \boldsymbol{B} = 0$ combined with the Lorentz force. When B reaches the value

$$B_{\mathrm{T}} = (v_0^2/v_{\perp 0}^2)B_0, \tag{2.37}$$

it follows from eqn (2.36) that this force has reduced v_z to zero.

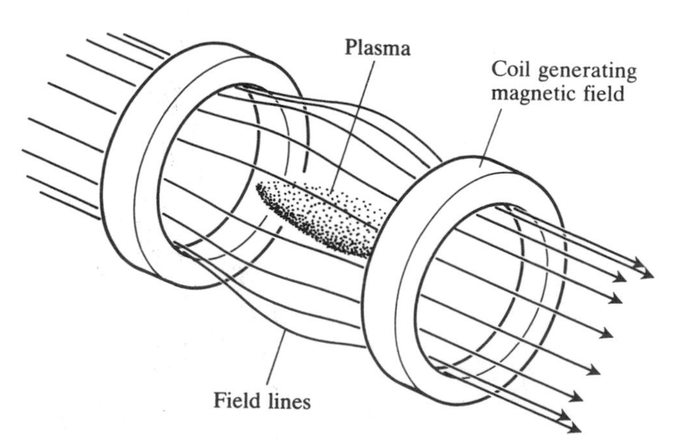

Fig. 2.7 Magnetic trapping.

Thereafter the particle reverses its path along the magnetic field. This is the phenomenon of magnetic trapping. Depending on their initial velocities, a certain proportion of the charged particles in an inhomogeneous magnetic field will be restricted by this mechanism to the region of space where the magnetic field is weakest. Magnetic trapping is the basis of all mirror plasma confinement systems (Fig. 2.7; compare Fig. 2.5).

Let the maximum and minimum values of the magnetic field in a mirror confinement system be B_M and B_0 respectively. It follows from eqn (2.37) that all particles whose initial velocities are such that $B_T \leqslant B_M$ will be trapped. All particles with a large initial ratio of parallel to perpendicular velocity, $v_{z0}^2/v_{\perp 0}^2 \geqslant (B_M - B_0)/B_0$, will escape. Mirror confinement systems are accordingly characterized by a loss cone distribution: that is, the region in (v_x, v_y, v_z)-space from which particles have escaped is formed of two cones, whose axes lie along the v_z-axis and whose tips touch at the origin, as in Fig. 2.8. It should be noted that particles which are initially trapped can undergo collisions which, by changing $v_{z0}^2/v_{\perp 0}^2$, scatter them into the loss cone so that they leave the system.

The trapping of electrons by a spatially varying magnetic field is significant in tokamak plasmas. In the absence of trapping, the guiding centre of an individual electron in a tokamak follows a magnetic field line, which is a helix wound around a toroidal surface centred on the magnetic axis, as shown in Fig. 2.9. The strength of the magnetic field at a given point is proportional to $1/R$, where R is the distance to the point from the axis of symmetry. It follows (see Fig. 2.10) that the magnetic field strength experienced by the electron at poloidal angle θ is proportional to $(R_0 + r_0 \cos\theta)^{-1}$. Here R_0 is the major radius of the

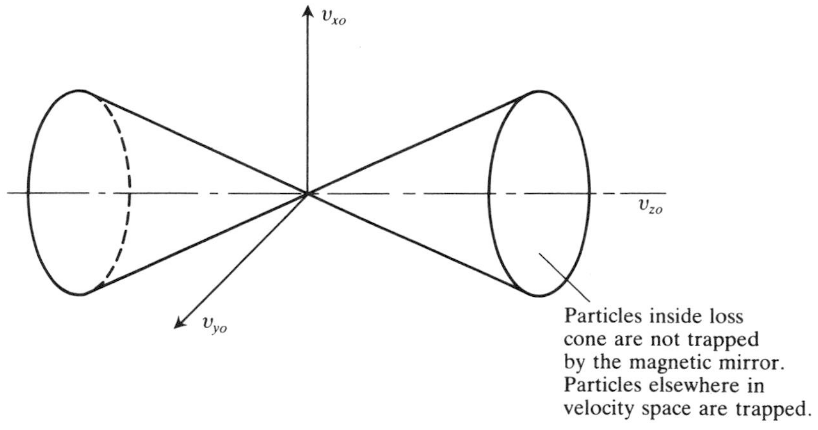

Fig. 2.8 Loss cone distribution.

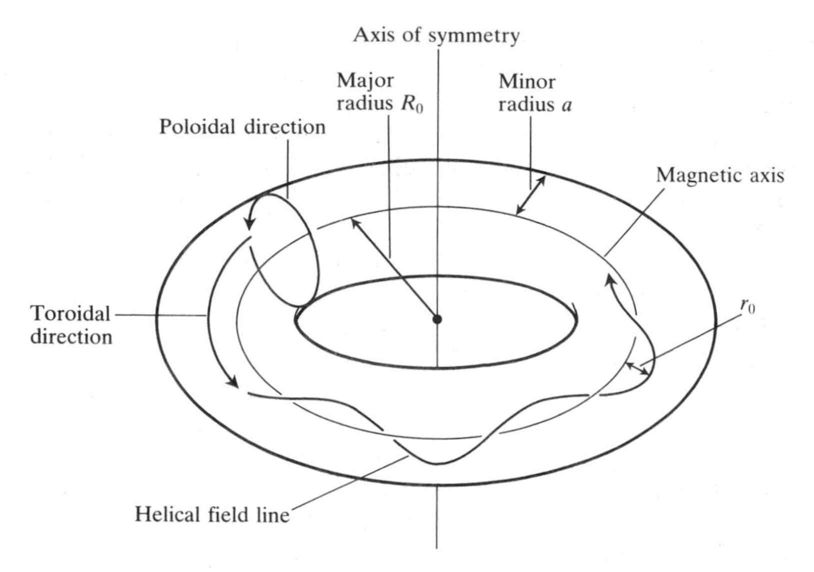

Fig. 2.9 Magnetic field line in a tokamak.

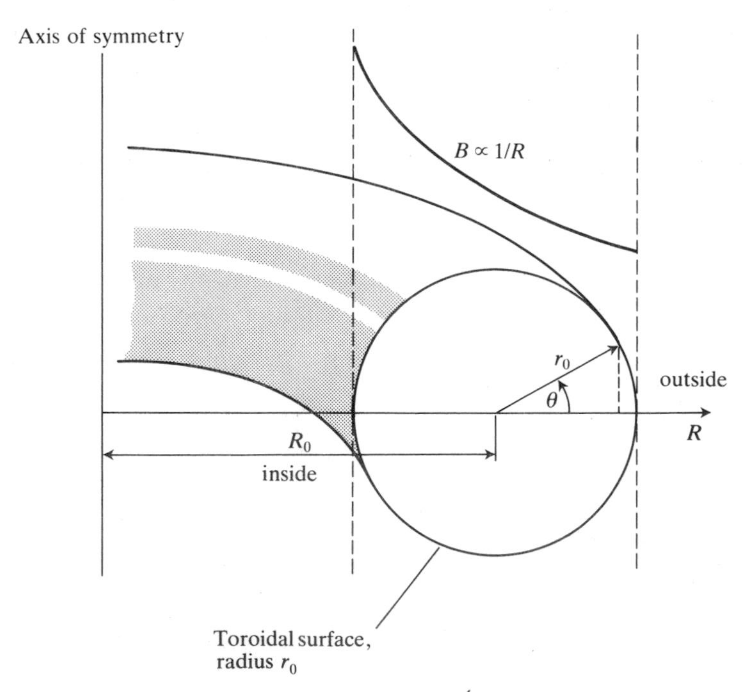

Fig. 2.10 Strength of magnetic field as a function of position on a poloidal cross-section of a tokamak.

tokamak; r_0 is the minor radius of the toroidal surface on which the guiding centre in question moves; and θ is zero on the outside of the toroidal surface. From our previous discussion, we expect some of the particles to undergo magnetic trapping in the inhomogeneous magnetic field that they experience. Such particles would be kept away from the region of greatest magnetic field strength, on the inside of the toroidal surface. The so-called inverse aspect ratio $\varepsilon = r_0/R_0$ is often a small parameter so that, using the binomial theorem, the magnetic field strength is

$$B(\theta) = B_0(1 - \varepsilon \cos \theta)/(1 - \varepsilon). \tag{2.38}$$

Here, the normalization is chosen so that B_0 is the field strength on the outside of the toroidal surface ($\theta = 0$), which is its minimum value on the surface. Substituting eqn (2.38) in eqn (2.36), we obtain

$$\frac{v_z^2}{v_0^2} = 1 - \frac{v_{\perp 0}^2}{v_0^2}\left(\frac{1 - \varepsilon \cos \theta}{1 - \varepsilon}\right). \tag{2.39}$$

It follows that it is possible for the toroidal velocity v_z to vanish before $\theta = \pi$, so that the electron cannot complete a poloidal circuit. By eqn (2.39), v_z becomes zero when the poloidal angle θ satisfies

$$\frac{v_{z0}^2}{v_{\perp 0}^2} = \varepsilon(1 - \cos \theta) \tag{2.40}$$

to leading order in ε. The value of the right-hand side of eqn (2.40) ranges between zero on the outside of the toroidal surface ($\theta = 0$) and 2ε on the inside ($\theta = \pi$). Then if

$$\frac{v_{z0}^2}{v_{\perp 0}^2} > 2\varepsilon, \tag{2.41}$$

eqn (2.40) cannot be satisfied and therefore v_z remains non-zero at all poloidal angles. These electrons are free to perform complete circuits, and are known as passing electrons; they correspond to the loss cone (Fig. 2.8). Conversely, if the ratio of parallel to perpendicular energy is so small that

$$\frac{v_{z0}^2}{v_{\perp 0}^2} < 2\varepsilon, \tag{2.42}$$

there is some poloidal angle θ_m at which eqn (2.40) is satisfied, and v_z vanishes there. These are the trapped electrons. The greater the degree to which $v_{z0}^2/v_{\perp 0}^2$ falls below 2ε, the more limited is the range of poloidal angles close to zero that the trapped electron explores. This corresponds, as expected, to confinement to the outside of the toroidal surface, which is the region of weakest magnetic field. Thus, the guiding centre of a trapped electron oscillates back and forth along the magnetic field line on the outside of the toroidal surface, between mirror points at $\pm\theta_m$.

It also follows from eqn (2.42) that the fraction of electrons on a given toroidal surface that are trapped is proportional to $\varepsilon^{\frac{1}{2}}$. Recall that $\varepsilon = r_0/R_0$. The value of R_0 is fixed, but r_0 can take any value between zero and the tokamak minor radius a, depending on the toroidal surface considered. Thus the proportion of trapped electrons increases outwards from the magnetic axis ($r_0 = 0$) in association with $\varepsilon^{\frac{1}{2}}$. In low aspect ratio tokamaks (large a/R_0), the variation in the magnetic field strength across the tokamak is proportionately large, and the fraction of electrons which this variation is sufficient to trap is correspondingly high. Finally we note that the motion of trapped electrons reverses periodically, so that they cannot contribute to the toroidal current which is fundamental to tokamak operation.

Exercises

2.1. Calculate the electron cyclotron frequency ω_{ce} for the following magnetic fields: (a) the earth's magnetic field near a pole, 6×10^{-5} T; (b) the galactic field, 3×10^{-10} T; (c) a sunspot, 0.25 T.

2.2. A particle with charge q is emitted from the origin with momentum p, directed at an angle θ to a uniform magnetic field B which lies in the z-direction. At what point does the particle next intersect the z-axis?

2.3. Consider a system of N non-interacting electrons in a uniform magnetic field. Suppose that they initially have an isotropic distribution of velocities, whose magnitude is v_0 for all the electrons. If the magnetic field strength increases adiabatically with time from B_1 to $B_2 = \alpha B_1$, with $\alpha > 1$, calculate the change in energy of the system.

Solutions are on pages 148 *and* 149.

3

Dielectric description of cold plasma

3.1 Introduction to the dielectric description of plasma

The macroscopic response of any medium to an applied electric field is determined by the sum of the microscopic responses of the individual particles that make up the medium. In a conducting medium, the macroscopic response is determined at a microscopic level by the separation between positive and negative charges that is produced by the applied electric field. If the applied field varies with time, so too will the microscopic state of the medium: the separation between the positive and negative charges will change with time, as will the electric field that is produced by their separation. Thus, particle currents and displacement currents are produced. At a macroscopic level, the particle current is described by the conductivity tensor* $\boldsymbol{\sigma}$, where

$$\boldsymbol{J} = \boldsymbol{\sigma} \cdot \boldsymbol{E}. \tag{3.1}$$

That is, $\boldsymbol{\sigma}$ is a macroscopic variable whose nature is determined by microscopic dynamics. Maxwell's equation eqn (I.9) states that $\boldsymbol{J}$ and the displacement current combine to act as the source of the magnetic field in the medium: $\nabla \times \boldsymbol{H} = \boldsymbol{J} + \varepsilon_0\, \partial \boldsymbol{E}/\partial t$. Using eqn (3.1), we have

$$\nabla \times \boldsymbol{H} = \left(\varepsilon_0 \frac{\partial}{\partial t} + \boldsymbol{\sigma}\cdot\right)\boldsymbol{E}. \tag{3.2}$$

Now recall eqns (I.10) and (I.11): $\boldsymbol{B} = \mu_0 \boldsymbol{H}$ and $\mu_0 \varepsilon_0 = 1/c^2$. Then if $\boldsymbol{E}$ varies as $\exp(-\mathrm{i}\omega t)$, eqn (3.2) gives

$$\nabla \times \boldsymbol{B} = -\frac{\mathrm{i}\omega}{c^2}\left(\boldsymbol{I} + \frac{\mathrm{i}}{\varepsilon_0 \omega}\boldsymbol{\sigma}\right) \cdot \boldsymbol{E} \tag{3.3}$$

where $\boldsymbol{I}$ is the identity matrix. It follows that all information about the macroscopic response of the medium to applied electric fields is contained in the dielectric tensor, defined by

$$\boldsymbol{\varepsilon} = \boldsymbol{I} + \frac{\mathrm{i}}{\varepsilon_0 \omega}\boldsymbol{\sigma}, \tag{3.4}$$

* For our purposes, $\boldsymbol{\sigma}$ is a 3×3 matrix. Thus, in eqn (3.1), $\boldsymbol{\sigma} \cdot \boldsymbol{E}$ represents the multiplication of the vector $\boldsymbol{E}$ by the matrix $\boldsymbol{\sigma}$.

and eqn (3.3) can now be written

$$\nabla \times \boldsymbol{B} = -\frac{i\omega}{c^2}\boldsymbol{\varepsilon} \cdot \boldsymbol{E}. \tag{3.5}$$

Operating on eqn (3.5) with $\partial/\partial t$, and using eqn (I.4) to eliminate $\partial\boldsymbol{B}/\partial t$, we obtain the wave equation

$$\nabla^2\boldsymbol{E} - \nabla(\nabla \cdot \boldsymbol{E}) + \frac{\omega^2}{c^2}\boldsymbol{\varepsilon} \cdot \boldsymbol{E} = 0. \tag{3.6}$$

So far, our discussion in this section has been completely general, and applies to any conducting medium. The nature of the macroscopic quantity $\boldsymbol{\sigma}$, and hence $\boldsymbol{\varepsilon}$, for a plasma is determined at the microscopic level by the plasma particle dynamics that we have described in the preceding sections. As the simplest example, consider eqn (1.1) which describes the motion of a plasma electron in the absence of an external magnetic field. It follows from eqn (1.1) that at the microscopic level, with an applied field $\boldsymbol{E}$ varying as $\exp(-i\omega t)$, each plasma electron will respond with a velocity

$$\boldsymbol{v} = (e/i\omega m)\boldsymbol{E}. \tag{3.7}$$

The current density associated with n_0 such electrons per unit volume is

$$\boldsymbol{J} = -n_0 e\boldsymbol{v} = -(n_0 e^2/i\omega m)\boldsymbol{E}, \tag{3.8}$$

so that $\boldsymbol{\sigma} = -(n_0 e^2/i\omega m)\boldsymbol{I}$. Then the definition eqn (3.4) of $\boldsymbol{\varepsilon}$ gives

$$\boldsymbol{\varepsilon} = (1 - \omega_{pe}^2/\omega^2)\boldsymbol{I}, \tag{3.9}$$

where we have used eqn (1.6).

Now let us calculate the normal modes—a macroscopic concept—of an unmagnetized plasma. If $\boldsymbol{E}$ varies as $\exp(i\boldsymbol{k} \cdot \boldsymbol{r} - i\omega t)$, eqns (3.6) and (3.9) combine to give

$$(\omega^2 - \omega_{pe}^2 - c^2k^2)\boldsymbol{E} + c^2\boldsymbol{k}(\boldsymbol{k} \cdot \boldsymbol{E}) = 0. \tag{3.10}$$

There are two classes of normal mode. First, consider the case of transverse modes, that have $\boldsymbol{k} \cdot \boldsymbol{E} = 0$ and are accordingly electromagnetic. Then eqn (3.10) gives

$$(\omega^2 - \omega_{pe}^2 - c^2k^2)\boldsymbol{E} = 0. \tag{3.11}$$

This is compatible with non-zero $\boldsymbol{E}$ only if ω and k are related by

$$\omega^2 = \omega_{pe}^2 + c^2k^2, \tag{3.12}$$

which is our first derivation of a dispersion relation. It tells us that the frequency of any electromagnetic wave in an unmagnetized plasma must exceed the electron plasma frequency. In addition, if we try to launch an electromagnetic wave into the plasma with frequency $\omega < \omega_{pe}$, eqn (3.12) indicates that the wave will have an imaginary wavenumber k

inside the plasma. The wave will therefore be evanescent, and unable to propagate through the plasma. Thus ω_{pe} plays the role of a cut-off frequency for electromagnetic waves in an unmagnetized plasma. This fact has a number of practical implications. For example, it determines the range of frequencies that can be used for different types of radiocommunication. In order to communicate with a satellite, one must choose a frequency that exceeds the plasma frequency of the ionospheric plasma. Otherwise, the signal will be reflected from the ionosphere, and will not reach the satellite. Conversely, we may send a radio signal to a distant point on the Earth's surface by choosing a frequency below the ionospheric plasma frequency, and use the ionosphere to reflect the signal in the required direction.

The second class of normal mode that satisfies eqn (3.10) is electrostatic, with $\boldsymbol{k}$ and $\boldsymbol{E}$ parallel. In this case, the dispersion relation is clearly

$$\omega^2 = \omega_{pe}^2. \tag{3.13}$$

Thus, our approach here confirms the result of Section 1.1, that the electrostatic normal modes of a plasma oscillate at the electron plasma frequency.

We note that the sophistication of the description provided by $\boldsymbol{\varepsilon}$ depends entirely on the sophistication of the model of the microscopic dynamics that has been employed in deriving it. In this section, we have used a model where the electrons are cold, that is, they have no random thermal motion, and the ions are stationary. All the relevant dynamics is contained in eqn (1.1). As we shall see in the chapter on kinetic theory, a more realistic microscopic model yields a more complicated form for $\boldsymbol{\varepsilon}$, which we shall use to describe a whole range of new phenomena. Meanwhile, we shall continue our development of plasma dielectric theory by turning to magnetized plasmas.

3.2 Dielectric tensor of a cold magnetized plasma

The effect on particle dynamics of the magnetic component of an electromagnetic wave can be shown to be negligible so long as the particle velocities are non-relativistic and $\omega \neq \omega_{ce}$. In this case, we need only consider the effect of the rapidly oscillating electric field. The motion of an electron parallel to the static uniform magnetic field is governed by

$$\dot{v}_z = -\frac{e}{m} E_z(t) \tag{3.14}$$

which is the same as eqn (1.1) because the magnetic field does not affect motion parallel to it. The perpendicular equation of motion,

$$\dot{\boldsymbol{v}}_\perp = -\frac{e}{m}\boldsymbol{E}_\perp(t) + \boldsymbol{\omega}_{\text{ce}} \times \boldsymbol{v}_\perp(t), \tag{3.15}$$

follows as usual from the Lorentz force eqn (I.16); compare eqn (2.3). Differentiating eqn (3.15) with respect to time, and substituting from eqn (3.15) for $\boldsymbol{\omega}_{\text{ce}} \times \dot{\boldsymbol{v}}_\perp$, we obtain

$$\ddot{\boldsymbol{v}}_\perp + \omega_{\text{ce}}^2 \boldsymbol{v}_\perp = -\frac{e}{m}(\dot{\boldsymbol{E}}_\perp + \boldsymbol{\omega}_{\text{ce}} \times \boldsymbol{E}_\perp). \tag{3.16}$$

In the absence of a perpendicular electric field, eqn (3.16) reduces to eqn (2.7), whose solution we shall denote by $\boldsymbol{v}_{\perp 0}(t)$; this describes cyclotron motion. Let us write $\boldsymbol{E}_\perp(t) = Re(\tilde{\boldsymbol{E}}_\perp \mathrm{e}^{-\mathrm{i}\omega t})$ and $\boldsymbol{v}_\perp(t) = \boldsymbol{v}_{\perp 0}(t) + Re(\tilde{\boldsymbol{v}}_\perp \mathrm{e}^{-\mathrm{i}\omega t})$. Then eqn (3.16) becomes

$$(\omega_{\text{ce}}^2 - \omega^2)\tilde{\boldsymbol{v}}_\perp = -\frac{e}{m}(-\mathrm{i}\omega\tilde{\boldsymbol{E}}_\perp + \boldsymbol{\omega}_{\text{ce}} \times \tilde{\boldsymbol{E}}_\perp). \tag{3.17}$$

Combining eqns (3.14) and (3.17),

$$\boldsymbol{v}(t) = \boldsymbol{v}_{\perp 0}(t) + Re(\boldsymbol{a} \cdot \tilde{\boldsymbol{E}} \mathrm{e}^{-\mathrm{i}\omega t}), \tag{3.18}$$

$$\boldsymbol{a} = \frac{e}{m}\begin{bmatrix} \dfrac{\mathrm{i}\omega}{\omega_{\text{ce}}^2 - \omega^2} & \dfrac{\omega_{\text{ce}}}{\omega_{\text{ce}}^2 - \omega^2} & 0 \\ \dfrac{-\omega_{\text{ce}}}{\omega_{\text{ce}}^2 - \omega^2} & \dfrac{\mathrm{i}\omega}{\omega_{\text{ce}}^2 - \omega^2} & 0 \\ 0 & 0 & -\mathrm{i}/\omega \end{bmatrix}. \tag{3.19}$$

All information about the particle drift arising from $\boldsymbol{E}_\perp$ is contained in $\boldsymbol{a}$. Setting $E_y = 0$, we see that

$$\tilde{v}_{\perp x}^2 + \tilde{v}_{\perp y}^2\left(\frac{\omega}{\omega_{\text{ce}}}\right)^2 = \left(\frac{e}{m}\right)^2 \frac{\omega^2}{\omega_{\text{ce}}^2 - \omega^2} E_x^2 \tag{3.20}$$

so that particle motion in the xy plane follows an ellipse, whose mean radius and eccentricity depend on the size of the electric field and the value of $\omega/\omega_{\text{ce}}$. This motion is superimposed on oscillation along the magnetic field in the harmonic field E_z.

In eqn (3.18), the response of the electron to the wave field is given by the second term on the right. This is the term on which we wish to concentrate, and we shall accordingly adopt a model in which the plasma is cold, so that the electrons have no random thermal motion. In terms of eqns (2.8) and (2.10), $v_\perp = r_{\text{L}} = 0$ and consequently in eqn (3.18), $v_{\perp 0}(t) = 0$. Then the current $\boldsymbol{j}(t)$ associated with the response to an electromagnetic wave of the n_0 electrons contained in unit volume of magnetized plasma is

$$\boldsymbol{j}(t) = -n_0 e \boldsymbol{v}(t) = -n_0 e \boldsymbol{a} \cdot \boldsymbol{E}(t) = \boldsymbol{\sigma} \cdot \boldsymbol{E}(t). \tag{3.21}$$

This defines the conductivity tensor $\boldsymbol{\sigma}$ for a cold, magnetized plasma. We now obtain the magnetic field which is set up in the plasma by the oscillating electric field and by the response to it of the electrons from eqn (3.5). The dielectric tensor $\boldsymbol{\varepsilon}$ in eqn (3.5) for a cold magnetized plasma follows from eqns (3.4), (3.21), and (3.19):

$$\boldsymbol{\varepsilon} = \begin{bmatrix} 1 + \dfrac{\omega_{pe}^2}{\omega_{ce}^2 - \omega^2} & \dfrac{-i\omega_{ce}}{\omega} \dfrac{\omega_{pe}^2}{\omega_{ce}^2 - \omega^2} & 0 \\ \dfrac{i\omega_{ce}}{\omega} \dfrac{\omega_{pe}^2}{\omega_{ce}^2 - \omega^2} & 1 + \dfrac{\omega_{pe}^2}{\omega_{ce}^2 - \omega^2} & 0 \\ 0 & 0 & 1 - \dfrac{\omega_{pe}^2}{\omega^2} \end{bmatrix}. \tag{3.22}$$

We have thus calculated the cold plasma dielectric tensor $\boldsymbol{\varepsilon}$ from considerations of particle dynamics, represented by eqns (3.14) and (3.15). The tensor $\boldsymbol{\varepsilon}$ contains a large amount of information. In particular, it governs the propagation of electromagnetic waves in magnetized plasmas.

Let us look for wave fields which oscillate as $\exp(i\boldsymbol{k} \cdot \boldsymbol{r} - i\omega t)$ and propagate in cold magnetized plasma. In this case, the general expression eqn (3.6) can be written

$$\boldsymbol{M} \cdot \boldsymbol{E} = 0, \tag{3.23}$$

where

$$\boldsymbol{M} = \frac{\omega^2}{c^2} \boldsymbol{\varepsilon} + \boldsymbol{kk} - k^2 \boldsymbol{I} \tag{3.24}$$

and $\boldsymbol{I}$ is the identity matrix. The normal modes of the system are accordingly given by

$$\det(\boldsymbol{M}) = 0. \tag{3.25}$$

Equation (3.25) is the basic dispersion relation: the wavevector $\boldsymbol{k}$ and frequency ω are related to each other by the components of the dielectric tensor $\boldsymbol{\varepsilon}$, which themselves contain ω and the parameters ω_{pe} and ω_{ce} which describe the state of the plasma. The solutions of eqn (3.25) describe the electromagnetic waves that propagate in a cold magnetized plasma. It is remarkable that all this information can be obtained using only the single-particle dynamics embodied in eqns (3.14) and (3.15), with the collective behaviour described by the simple summation over all electrons in eqn (3.21).

3.3 High-frequency waves in a cold magnetized plasma

Let us now consider the high-frequency normal modes of a cold magnetized plasma. The frequency of each normal mode is a root of eqn (3.25); the corresponding polarization is obtained from eqn (3.23). To help with the mainpulation of the dielectric tensor $\boldsymbol{\varepsilon}$ given by eqn (3.22), we define

$$\varepsilon_1 = 1 + \frac{\omega_{\mathrm{pe}}^2}{\omega_{\mathrm{ce}}^2 - \omega^2}, \tag{3.26}$$

$$\varepsilon_2 = \frac{\omega_{\mathrm{ce}}}{\omega} \frac{\omega_{\mathrm{pe}}^2}{\omega_{\mathrm{ce}}^2 - \omega^2}, \tag{3.27}$$

$$\varepsilon_3 = 1 - \frac{\omega_{\mathrm{pe}}^2}{\omega^2}, \tag{3.28}$$

so that

$$\boldsymbol{\varepsilon} = \begin{bmatrix} \varepsilon_1 & -\mathrm{i}\varepsilon_2 & 0 \\ \mathrm{i}\varepsilon_2 & \varepsilon_1 & 0 \\ 0 & 0 & \varepsilon_3 \end{bmatrix}. \tag{3.29}$$

Note that ω_{ce}, defined in eqn (2.4), is a positive quantity. We shall also find it useful to define a vector $\boldsymbol{N}$, as follows, whose magnitude is given by the refractive index and whose direction is given by the wavevector $\boldsymbol{k}$. Thus,

$$\boldsymbol{N} = c\boldsymbol{k}/\omega. \tag{3.30}$$

Then we may write $\boldsymbol{M}$, defined in eqn (3.24), as

$$\boldsymbol{M} = \frac{\omega^2}{c^2} \begin{bmatrix} \varepsilon_1 - N_y^2 - N_z^2 & -\mathrm{i}\varepsilon_2 + N_x N_y & N_x N_z \\ \mathrm{i}\varepsilon_2 + N_x N_y & \varepsilon_1 - N_x^2 - N_z^2 & N_y N_z \\ N_x N_z & N_y N_z & \varepsilon_3 - N_x^2 - N_y^2 \end{bmatrix}. \tag{3.31}$$

Consider first waves that propagate parallel to the magnetic field. In this case, $N_x = N_y = 0$, and $N_z = N$. By eqn (3.31), $\boldsymbol{M}$ reduces to

$$\boldsymbol{M} = \frac{\omega^2}{c^2} \begin{bmatrix} \varepsilon_1 - N^2 & -\mathrm{i}\varepsilon_2 & 0 \\ \mathrm{i}\varepsilon_2 & \varepsilon_1 - N^2 & 0 \\ 0 & 0 & \varepsilon_3 \end{bmatrix}. \tag{3.32}$$

The corresponding normal modes of the plasma have frequencies that are the roots of $\det(\boldsymbol{M}) = 0$, that is

$$\varepsilon_3\{(\varepsilon_1 - N^2)^2 - \varepsilon_2^2\} = 0. \tag{3.33}$$

This equation is satisfied first by

$$\varepsilon_3 = 0 \tag{3.34}$$

with $(\varepsilon_1 - N^2)^2 - \varepsilon_2^2 \neq 0$. Then by the definition eqn (3.28) of ε_3, eqn (3.34) gives

$$\omega^2 = \omega_{\mathrm{pe}}^2. \tag{3.35}$$

To establish the polarization of this wave, substitute eqn (3.34) into eqn (3.32). In this case, eqn (3.23) will be satisfied by an electric field of the form $\boldsymbol{E} = (0, 0, E_z)$. Thus, this normal mode of a cold magnetized plasma is an electrostatic wave, whose wavevector and field amplitude are both directed along the magnetic field. The wave oscillates at the plasma frequency, and is indistinguishable from the Langmuir wave in a plasma with no magnetic field, that we met in Chapter 1. Physically, each electron is acted on by an electric field that is parallel to the magnetic field, so that there is no perpendicular component of motion that could be affected by the Lorentz force: $\boldsymbol{v} \times \boldsymbol{B} = 0$.

The other normal modes that propagate parallel to the magnetic field follow from eqn (3.33) with $\varepsilon_3 \neq 0$ and

$$\varepsilon_1 - N^2 = \pm \varepsilon_2. \tag{3.36}$$

Let us consider first the case

$$\varepsilon_1 - \varepsilon_2 = N^2. \tag{3.37}$$

Using eqns (3.26) and (3.27), this gives the following dispersion relation between k $(= \omega N/c)$ and ω:

$$k = \frac{\omega}{c}\left\{1 - \frac{\omega_{\mathrm{pe}}^2}{\omega(\omega_{\mathrm{ce}} + \omega)}\right\}^{\frac{1}{2}}. \tag{3.38}$$

The polarization of this wave is obtained by substituting eqn (3.37) into eqn (3.32), and setting $\boldsymbol{M} \cdot \boldsymbol{E} = 0$. This is satisfied by an electric field of the form $\boldsymbol{E} = (E_x, E_y, 0)$, where

$$E_x - \mathrm{i}E_y = 0. \tag{3.39}$$

It follows from eqn (3.39) that when $E_x \sim \exp(\mathrm{i}k_z - \mathrm{i}\omega t)$, we must have $E_y \sim \exp(\mathrm{i}kz - \mathrm{i}\omega t + \mathrm{i}\pi/2)$. Taking real parts, and denoting the phase $kz - \omega t$ by $\phi(t)$, we have $E_x \sim \cos\phi(t)$ and $E_y \sim \cos(\phi(t) + \pi/2)$. Thus when $\phi = 0$, $\boldsymbol{E}$ lies entirely in the positive x-direction. As ϕ increases from zero, E_x decreases in magnitude and E_y grows in the negative y-direction. That is, $\boldsymbol{E}$ rotates in the xy-plane in the direction of the curled fingers of the left hand when the thumb lies along the magnetic field, which as usual defines the z-direction. Therefore eqn (3.38) is the dispersion relation of a left circularly polarized wave.

Returning to eqn (3.36), let us consider the second case

$$\varepsilon_1 + \varepsilon_2 = N^2. \tag{3.40}$$

Again using eqns (3.26) and (3.27), this gives the dispersion relation

$$k = \frac{\omega}{c}\left\{1 + \frac{\omega_{\mathrm{pe}}^2}{\omega(\omega_{\mathrm{ce}} - \omega)}\right\}^{\frac{1}{2}}. \tag{3.41}$$

Substituting eqn (3.40) into eqn (3.32), and considering $\boldsymbol{M} \cdot \boldsymbol{E} = 0$, it follows that

$$E_x + iE_y = 0. \tag{3.42}$$

By repeating the phase analysis of the preceding paragraph, it can be seen that this is a right circularly polarized wave. We note from eqns (2.2) and (2.3) that this sense of rotation is the same as that of an electron in the magnetic field. Conversely, as we shall now see, the ions, which are oppositely charged, rotate in a left-handed sense about the magnetic field. For an ion which has mass M and charge Ze, the Lorentz force equation is

$$M\dot{\boldsymbol{v}} = Ze\boldsymbol{v} \times \boldsymbol{B}. \tag{3.43}$$

It is useful to define a vector

$$\boldsymbol{\omega}_{\text{ci}} = -\frac{Ze\boldsymbol{B}}{M}, \tag{3.44}$$

which is oppositely directed to $\boldsymbol{\omega}_{\text{ce}}$ defined by eqn (2.2). Then eqn (3.43) becomes

$$\dot{\boldsymbol{v}} = \boldsymbol{\omega}_{\text{ci}} \times \boldsymbol{v}, \tag{3.45}$$

corresponding to rotation in the opposite sense to the electrons. The magnitude of $\boldsymbol{\omega}_{\text{ci}}$ is the ion cyclotron frequency

$$\omega_{\text{ci}} = \frac{ZeB}{M} = 9.6 \times 10^7\, Z\left(\frac{B}{1\ \text{Tesla}}\right)\left(\frac{M}{m_{\text{p}}}\right)^{-1} \text{rad s}^{-1}. \tag{3.46}$$

where m_{p} denotes the proton mass. We note that even for the lightest ion, a proton, ω_{ci} is smaller than ω_{ce} by the factor $m_{\text{p}}/m_{\text{e}}$—three orders of magnitude. It is for this reason that we could afford to neglect ion dynamics in our consideration of high-frequency normal modes, with ω greater than ω_{pe} and ω_{ce}. We shall quantify the range of validity of eqns (3.38) and (3.41) in greater detail in the next section, when we return to the topic of ion dynamics.

Noting that ion dynamics are neglected, let us consider the left circularly polarized dispersion relation, eqn (3.38). We note first that when ω takes the value

$$\omega_1 = \frac{\omega_{\text{ce}}}{2}\{\surd(1 + 4\omega_{\text{pe}}^2/\omega_{\text{ce}}^2) - 1\}, \tag{3.47}$$

the magnitude k of the wavevector is zero, corresponding to infinite wavelength. Equation (3.38) has no real (as opposed to imaginary) solution for k when $\omega < \omega_1$, so that ω_1 is known as the cut-off frequency. If the plasma parameters are such that $\omega_{\text{pe}}^2/\omega_{\text{ce}}^2 < 2$, it follows from eqn (3.47) that $\omega_1 < \omega_{\text{ce}}$; and vice versa. At high frequencies $\omega \gg \omega_{\text{pe}}$ and ω_{ce}, eqn (3.38) gives $\omega \simeq ck$, so that the waves propagate with a phase velocity, close to the speed of light in vacuum, which is almost independent of k. These waves are unaffected by the presence of the plasma for two related reasons. First, the plasma particles are too heavy to be able to respond coherently to the rapidly changing wave field. Second, the timescale of electron cyclotron motion is much longer than that of the wave oscillation. The electrons hardly move on their Larmor orbits while the wave phase reverses, so that the electrons are effectively unmagnetized with respect to the wave.

The dispersion relation eqn (3.41) for right circularly polarized waves has a more complicated structure than eqn (3.38). We note first that as ω approaches ω_{ce} from below, the magnitude k of the wavevector tends to infinity, corresponding to zero wavelength. This phenomenon occurs in general when the normal mode corresponds to some resonance of the system. In this case, it reflects the equality of the frequency of the right circularly polarized wave with the frequency ω_{ce} of electron gyration in the magnetic field, which is also a right-handed motion. The analogous resonance for left circularly polarized waves occurs when $\omega = \omega_{ci}$; as we have seen, this frequency lies beyond the range of validity of eqn (3.38). Returning to eqn (3.41), we see that if ω exceeds ω_{ce} by a small amount, the quantity $\omega_{pe}^2/\omega(\omega_{ce} - \omega)$ is negative and has a magnitude much greater than unity, so that no real value of k satisfies the equation. As ω is increased further, the magnitude of $\omega_{pe}^2/\omega(\omega_{ce} - \omega)$ is reduced, and when ω reaches the value

$$\omega_2 = \frac{\omega_{ce}}{2}\{\sqrt{(1 + 4\omega_{pe}^2/\omega_{ce}^2)} + 1\}, \tag{3.48}$$

eqn (3.41) has the solution $k = 0$. For $\omega > \omega_2$, eqn (3.41) again has real solutions. Thus ω_2 is the low-frequency cutoff of a second branch of right circularly polarized waves. At very high frequencies $\omega \gg \omega_{pe}$ and ω_{ce}, this second branch has $\omega \simeq ck$, which repeats the behaviour that we noted for high-frequency left circularly polarized waves.

Let us now summarize the properties of the normal modes of a cold magnetized plasma that have $\boldsymbol{k}$ parallel to the magnetic field direction and frequency ω significantly greater than ω_{ci}. Right circularly polarized waves exist for $\omega < \omega_{ce}$ and $\omega > \omega_2$. The branch with $\omega < \omega_{ce}$, which has a resonance at $\omega = \omega_{ce}$, is known as the Whistler wave.* By eqn (3.41), when ω is sufficiently small (though still large compared to characteristic ion frequencies) that $\omega_{ce} \gg \omega$ and $\omega_{pe}^2/\omega_{ce}\omega \gg 1$, the Whistler wave frequency is given approximately by

$$\omega = (ck/\omega_{pe})^2\omega_{ce}. \tag{3.49}$$

* So-called because of the audio characteristics of these waves, which were first picked up by military signalling equipment during the First World War; see Exercise 3.2. The waves originated in lightning discharges, and travelled to ground along the lines of the Earth's magnetic field. Interestingly, there is at present much research on the effects of pulses of electromagnetic radiation (generated by thermonuclear explosions or, in the laboratory, by plasma sources) on communications equipment.

A full discussion of this branch must await the inclusion of ion dynamics in the next section. The plasma has no right circularly polarized normal modes in the frequency range $\omega_{ce} < \omega < \omega_2$. A left circularly polarized normal mode exists when $\omega > \omega_1$. If $\omega_{pe}^2/\omega_{ce}^2 > 2$, it follows from eqn (3.47) that $\omega_1 > \omega_{ce}$; there is then a range of frequencies $\omega_{ce} < \omega < \omega_1$ for which the plasma has no electromagnetic normal modes of either circular polarization. Finally, there is the electrostatic normal mode with $\omega = \omega_{pe}$. The relation between k and ω for all these normal modes is illustrated in Fig. 3.1 for the case $\omega_{pe}^2/\omega_{ce}^2 < 2$, so that $\omega_1 < \omega_{ce}$.

We shall now consider the normal modes that have wavevector $\boldsymbol{k}$ perpendicular to the magnetic field direction. In the homogeneous magnetic field that we consider, the magnetic field properties remain the same as we move in any perpendicular direction. It follows that the designation of any particular perpendicular direction as the x-axis is arbitrary, because of the rotational symmetry of the system about the magnetic field direction. Without loss of generality, we may therefore choose the x-axis to lie in the direction parallel to the wavevector $\boldsymbol{k}$. Then by eqn (3.30), $N_y = N_z = 0$, and eqn ((3.31) becomes

$$\boldsymbol{M} = \begin{bmatrix} \varepsilon_1 & -\mathrm{i}\varepsilon_2 & 0 \\ \mathrm{i}\varepsilon_2 & \varepsilon_1 - N^2 & 0 \\ 0 & 0 & \varepsilon_3 - N^2 \end{bmatrix}. \tag{3.50}$$

The corresponding normal modes of the plasma have frequencies that

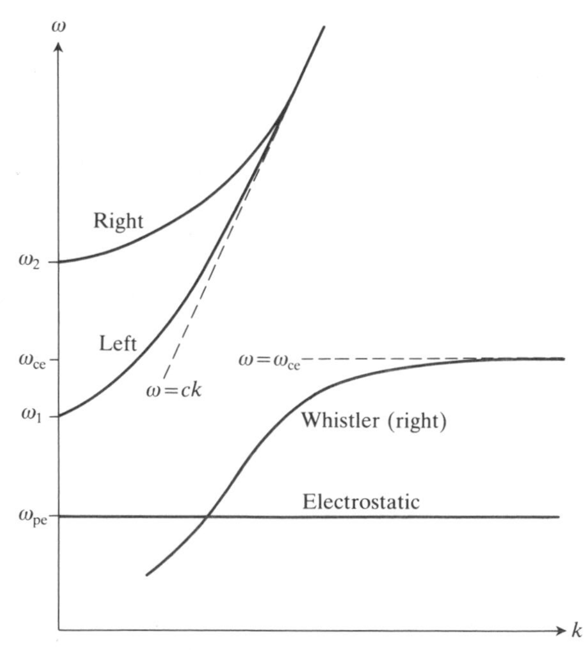

Fig. 3.1 Relation between ω and k for waves that propagate parallel to the magnetic field.

are, as usual, the roots of det($\boldsymbol{M}$) = 0, which now gives

$$(\varepsilon_3 - N^2)\{\varepsilon_1(\varepsilon_1 - N^2) - \varepsilon_2^2\} = 0. \tag{3.51}$$

This equation is satisfied first by

$$\varepsilon_3 = N^2 \tag{3.52}$$

with $\varepsilon_1(\varepsilon_1 - N^2) - \varepsilon_2^2 \neq 0$. By the definition of ε_3 in eqn (3.28), eqn (3.52) gives

$$\omega^2 = \omega_{\text{pe}}^2 + c^2k^2 \tag{3.53}$$

which is identical to the dispersion relation eqn (3.11) for electromagnetic waves propagating in arbitrary directions in an unmagnetized plasma. The fact that the plasma is magnetized does not enter eqn (3.53), and this normal mode is accordingly referred to as the Ordinary mode (O-mode). Physically, the non-appearance of the magnetic field indicates that the particle dynamics associated with the O-mode must take place exclusively in the direction parallel to the magnetic field, so that $\boldsymbol{v} \times \boldsymbol{B} = 0$ in the Lorentz force. Therefore the O-mode must be polarized with an amplitude $(0, 0, E_z)$. This can be confirmed in the usual way, by substituting eqn (3.52) into eqn (3.50), and considering the equation $\boldsymbol{M} \cdot \boldsymbol{E} = 0$.

Equation (3.51) is also satisfied by

$$\varepsilon_1(\varepsilon_1 - N^2) - \varepsilon_2^2 = 0 \tag{3.54}$$

with $\varepsilon_3 - N^2 \neq 0$. Using eqns (3.26) and (3.27), this gives

$$k = \frac{1}{c}\left\{\frac{(\omega^2 - \omega_1^2)(\omega^2 - \omega_2^2)}{\omega^2 - \omega_{\text{UH}}^2}\right\}^{\frac{1}{2}}. \tag{3.55}$$

Here ω_1 and ω_2 are the frequencies that were previously defined in eqns (3.47) and (3.48) respectively, and ω_{UH} denotes the upper hybrid frequency:

$$\omega_{\text{UH}} = (\omega_{\text{pe}}^2 + \omega_{\text{ce}}^2)^{\frac{1}{2}}. \tag{3.56}$$

The term hybrid is used because this frequency combines two distinct aspects of electron plasma dynamics: space-charge density oscillation (ω_{pe}) and electron cyclotron motion (ω_{ce}). Let us first consider the magnitudes of ω_1 and ω_2 relative to ω_{UH}. Combining eqns (3.47) and (3.56),

$$\omega_1 = (\omega_{\text{UH}}^2 - 3\omega_{\text{ce}}^2/4)^{\frac{1}{2}} - \omega_{\text{ce}}/2 \tag{3.57}$$

so that ω_1 is always less than ω_{UH}, irrespective of the values of ω_{pe} and ω_{ce}. Combining eqns (3.49) and (3.56),

$$\omega_2 = \frac{\omega_{\text{UH}}}{2}\{(1 + 3\omega_{\text{pe}}^2/\omega_{\text{UH}}^2)^{\frac{1}{2}} + (1 - \omega_{\text{pe}}^2/\omega_{\text{UH}}^2)^{\frac{1}{2}}\}\,. \tag{3.58}$$

The first square-root term in eqn (3.58) exceeds unity by more than the amount by which the second square-root term is less than unity. Thus the value of the entire term in { } in eqn (3.58) must exceed two, and ω_2 is always greater than ω_{UH}, irrespective of the values of ω_{pe} and ω_{ce}. So, quite generally,

$$\omega_1 < \omega_{\mathrm{UH}} < \omega_2. \tag{3.59}$$

Now let us return to eqn (3.55). It follows from eqn (3.59) that there are real solutions k only when $\omega_{\mathrm{UH}} > \omega > \omega_1$ or $\omega > \omega_2$. In addition, the value of k tends to infinity as ω approaches ω_{UH} from below. It follows that there are two regions of propagation for this wave, which is known as the Extraordinary mode (X-mode). The low-frequency branch has a low-frequency cut-off at $\omega = \omega_1$, and a resonance at $\omega = \omega_{\mathrm{UH}}$. The high-frequency branch has a low-frequency cut-off at $\omega = \omega_2$, and by eqn (3.55) satisfies $\omega \simeq ck$ at high frequencies, as expected. In Fig. 3.2, the frequencies of the ordinary mode and both branches of the Extraordinary mode are plotted as a function of k.

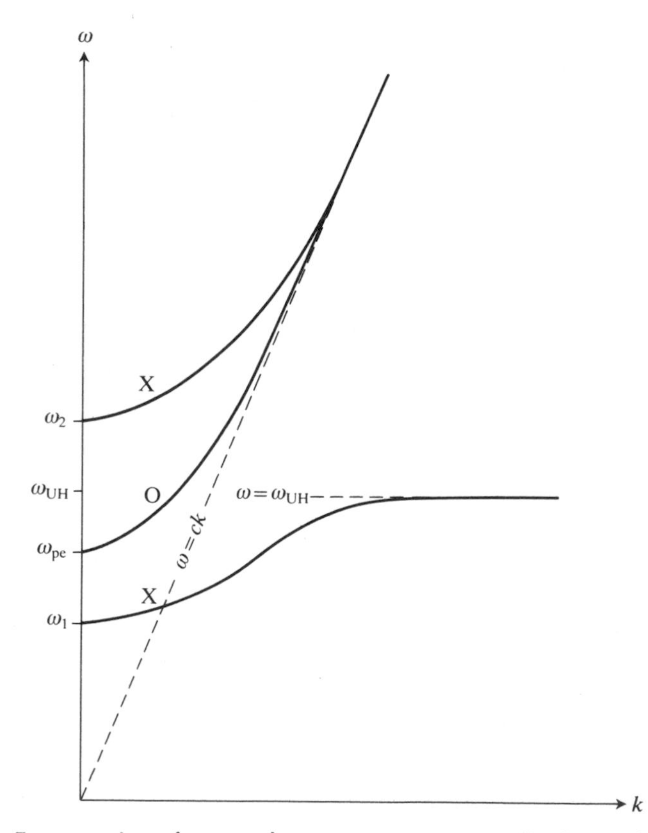

Fig. 3.2 Frequencies of waves that propagate perpendicular to the magnetic field, showing the Ordinary mode and both branches of the Extraordinary mode.

So far, we have considered the Extraordinary mode dispersion relation eqn (3.54) in the form eqn (3.55), which is convenient for establishing the relations between ω and k given fixed ω_{pe} and ω_{ce}, and hence fixed ω_1 and ω_2. It is also of interest to consider the reverse question. Given a plasma in which ω_{pe} and ω_{ce} vary with position, what regions of the plasma are accessible to an Extraordinary mode of fixed ω? Using the fact that, by eqns (3.47) and (3.48), $\omega_1^2 + \omega_2^2 = \omega_{ce}^2 + 2\omega_{pe}^2$ and $\omega_1^2\omega_2^2 = \omega_{pe}^4$, it is convenient to write the term appearing in the numerator of eqn (3.55) as

$$(\omega^2 - \omega_1^2)(\omega^2 - \omega_2^2) = (\omega_{pe}^2 - \omega^2)^2 - \omega_{ce}^2\omega^2. \tag{3.60}$$

Combining eqns (3.55), (3.56), and (3.60), we obtain for the Extraordinary mode

$$N_x^2 \equiv \left(\frac{ck}{\omega}\right)^2 = \frac{\left(\dfrac{\omega_{pe}^2}{\omega^2} - 1\right)^2 - \dfrac{\omega_{ce}^2}{\omega^2}}{1 - \left(\dfrac{\omega_{pe}^2}{\omega^2} + \dfrac{\omega_{ce}^2}{\omega^2}\right)}. \tag{3.61}$$

It is usual to define variables X and Y by

$$X = \omega_{pe}^2/\omega^2, \tag{3.62}$$

$$Y = \omega_{ce}/\omega. \tag{3.63}$$

These are chosen to be dimensionless variables whose magnitude is proportional to the physical parameters of interest. By eqn (1.6), X is proportional to the plasma density given fixed ω, and by eqn (2.4), Y is proportional to the magnetic field strength given fixed ω. In terms of these variables, the Extraordinary mode dispersion relation eqn (3.61) becomes

$$N_x^2 = \frac{(X-1)^2 - Y^2}{1 - X - Y^2}. \tag{3.64}$$

Also, the Ordinary mode dispersion relation eqn (3.53) becomes

$$N_o^2 = 1 - X. \tag{3.65}$$

This formulation makes it simple to recognize the mode cut-offs and resonances in terms of physical variables. By eqn (3.64), Extraordinary mode cut-offs $N_x = 0$ occur when $X - 1 = \pm Y$. Thus, for a given value of Y, there are two Extraordinary mode cut-offs; the one with the lower value of X (low-density cut-off) occurs at

$$X = 1 - Y, \tag{3.66}$$

and the one with the higher value of X (high-density cut-off) occurs at

$$X = 1 + Y. \tag{3.67}$$

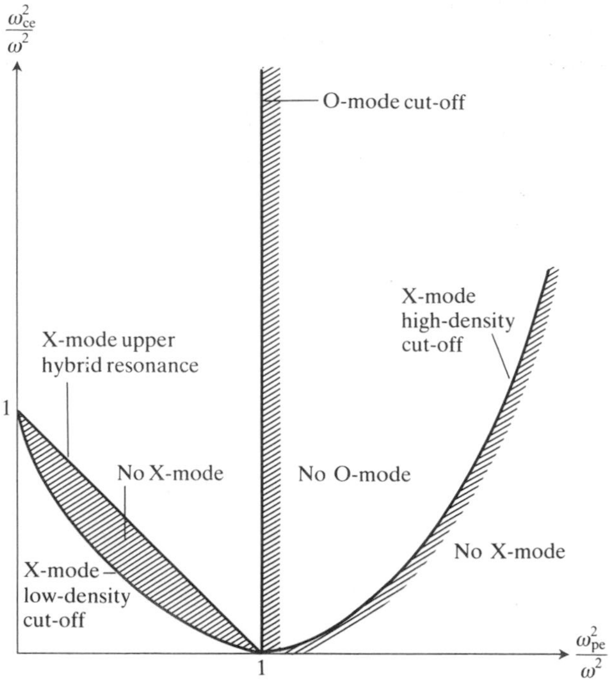

Fig. 3.3 Dispersion surfaces.

Note that the fact that we can write down eqns (3.66) and (3.67) does not necessarily mean that corresponding surfaces exist in the plasma. For example, if the density of the plasma is everywhere so low that the maximum value of X is less than the minimum value of $1 + Y$, there is no high-density cut-off present for the mode considered. Returning to eqn (3.64), the Extraordinary mode upper hybrid resonance occurs at

$$X + Y^2 = 1, \tag{3.68}$$

which by eqns (3.62) and (3.63) corresponds to eqn (3.56). Finally, by eqn (3.65), the Ordinary mode cut-off occurs at $X = 1$, and N_o is imaginary for $X > 1$ (which implies $\omega < \omega_{pe}$) as expected. In Fig. 3.3, these surfaces are plotted with $X = \omega_{pe}^2/\omega^2$ as the x-coordinate and $Y^2 = \omega_{ce}^2/\omega^2$ as the y-coordinate. This type of diagram is known as a CMA diagram, after its originators Clemmow, Mullaly, and Allis.

3.4 Low-frequency waves in cold magnetized plasma

In the preceding section, ion dynamics were explicitly neglected. Accordingly, we only considered normal modes with frequencies greater than or comparable to ω_{pe} and ω_{ce}. Let us now quantify the frequency regime in which ion dynamics become significant. We have already seen in eqn (3.46) that the ion cyclotron frequency ω_{ci} corresponds typically to a few tens of MHz. In addition to ion cyclotron motion, we must consider the effect of ion space charge, which is described by the ion plasma frequency ω_{pi}—the ion equivalent to ω_{pe}. As before, consider ions of charge Ze and mass M, in a plasma with overall charge neutrality so that the ion number density n_i satisfies

$$n_i = n_0/Z, \tag{3.69}$$

where n_0 is the electron number density. Using arguments analogous to those in Section 1.1, we see that following eqn (1.6), and using eqn (3.69), $\omega_{pi}^2 = (n_0/Z)(Ze)^2/\varepsilon_0 M$. Then

$$\omega_{pi} = (n_0 Ze^2/\varepsilon_0 M)^{\frac{1}{2}} \tag{3.70}$$

$$= 4.2 \times 10^9\, Z^{\frac{1}{2}} \left(\frac{n_0}{10^{19}\,\mathrm{m}^{-3}}\right)^{\frac{1}{2}} \left(\frac{M}{m_p}\right)^{-\frac{1}{2}} \mathrm{rad\,s}^{-1} \tag{3.71}$$

in an electrically neutral plasma. In a typical thermonuclear fusion experiment, ω_{pi} corresponds to frequencies of a few GHz. Comparing eqns (3.71) and (3.46), it follows that ω_{pi} typically exceeds ω_{ci} by at least an order of magnitude. In contrast, we noted after eqn (2.10) that the corresponding electron frequencies ω_{pe} and ω_{ce} are typically comparable to each other in magnitude. It is also useful to note that in an electrically neutral plasma, eqns (1.6), (2.4), (3.46) and (3.70) imply

$$\omega_{pe}^2/\omega_{ce} = \omega_{pi}^2/\omega_{ci}. \tag{3.72}$$

The contribution which the ions make to the dielectric response of the plasma is calculated in the same way as the electron contribution was obtained in Section 3.2. We can save ourselves the trouble of repeating these arguments, by substituting ions for electrons and making appropriate adjustments, as follows. First of all, the only quantities appearing in the expression for $\boldsymbol{\varepsilon}$ in eqn (3.22) that are specific to the electrons are ω_{ce} and ω_{pe}^2. Now ω_{ce} enters the electron dynamics through eqns (2.2) and (2.3). The corresponding equations for the ion dynamics are eqns (3.44) and (3.45). These pairs of equations are formally identical, but $\boldsymbol{\omega}_{ci}$ and $\boldsymbol{\omega}_{ce}$ are oppositely directed along the magnetic field direction, reflecting the opposite charge and opposite sense of gyration of electrons and ions. Thus, the cyclotron part of the transformation of eqn (3.22) consists of replacing ω_{ce} by $-\omega_{ci}$. The space charge part of the transformation is even simpler. In eqn (3.22), ω_{pe} enters only in the form ω_{pe}^2, and the squaring eliminates any possible dependence on the sign of the electron charge. We therefore complete the transformation by replacing ω_{pe}^2 by ω_{pi}^2. It follows that we may now write $\boldsymbol{\varepsilon}$ in the form of eqn (3.29), but instead of eqns (3.26) to (3.28) we have the following more general terms, where we have added together the effects of electron dynamics and ion dynamics:

$$\varepsilon_1 = 1 + \frac{\omega_{pe}^2}{\omega_{ce}^2 - \omega^2} + \frac{\omega_{pi}^2}{\omega_{ci}^2 - \omega^2}, \tag{3.73}$$

$$\varepsilon_2 = \frac{\omega_{ce}}{\omega}\frac{\omega_{pe}^2}{\omega_{ce}^2 - \omega^2} - \frac{\omega_{ci}}{\omega}\frac{\omega_{pi}^2}{\omega_{ci}^2 - \omega^2}, \tag{3.74}$$

$$\varepsilon_3 = 1 - \frac{\omega_{pe}^2}{\omega^2} - \frac{\omega_{pi}^2}{\omega^2}. \tag{3.75}$$

In order to study the low-frequency normal modes of cold magnetized plasma, we must substitute the new expressions, eqns (3.73) to (3.75), into the previous equations for parallel and perpendicular propagation, given by eqns (3.33) and (3.51) respectively. Let us first consider propagation parallel to the magnetic field, and start with the root given by eqn (3.34). Using eqn (3.75), $\varepsilon_3 = 0$ becomes

$$1 - \frac{\omega_{pe}^2}{\omega^2}(1 + \omega_{pi}^2/\omega_{pe}^2) = 0. \tag{3.76}$$

We know that $\omega_{pi}^2/\omega_{pe}^2 \ll 1$, so that the only change due to ion dynamics is a minor increase in the cut-off frequency. Physically, ion dynamics have little effect on this mode for two reasons. First, as we noted previously, this mode is not affected by the presence of the magnetic field, so that ion cyclotron motion is irrelevant to it. Second, the high mobility of the light electrons prevents the emergence of any low frequency wave branch associated with ω_{pi}. An electric field that oscillated slowly, with a frequency near ω_{pi}, would immediately be cancelled out by the rapid, short-circuiting response of the electrons before the ions could move.

Now let us consider the other parallel normal modes, described by eqn (3.36) with ε_1 and ε_2 given by eqns (3.73) and (3.74). First, the left circularly polarized waves described by eqn (3.37) now satisfy

$$k = \frac{\omega}{c}\left\{1 - \frac{\omega_{pe}^2}{\omega(\omega_{ce} + \omega)} + \frac{\omega_{pi}^2}{\omega(\omega_{ci} - \omega)}\right\}^{\frac{1}{2}}. \tag{3.77}$$

Using the fact that, by eqn (3.72), $\omega_{pi}^2(\omega_{ce} + \omega) - \omega_{pe}^2(\omega_{ci} - \omega) = \omega(\omega_{pe}^2 + \omega_{pi}^2)$, eqn (3.77) can be written

$$k = \frac{\omega}{c}\left\{1 + \frac{\omega_{pe}^2 + \omega_{pi}^2}{(\omega_{ce} + \omega)(\omega_{ci} - \omega)}\right\}^{\frac{1}{2}}. \tag{3.78}$$

This shows a new left circularly polarized branch; it exists for $0 < \omega < \omega_{ci}$, and is called the ion cyclotron wave. There is a resonance at $\omega = \omega_{ci}$, because the ion sense of gyration in the magnetic field is also left-handed, as we noted at eqn (3.43). At frequencies just above ω_{ci}, there is no new left circularly polarized wave. This is because the corresponding small negative value of the term $(\omega_{ci} - \omega)$ in eqn (3.78) prevents the existence of real roots. When ω is significantly above ω_{ci}, the term $(\omega_{ci} - \omega)$ tends to $-\omega$, and since $\omega_{pi}^2 \ll \omega_{pe}^2$, eqn (3.78) tends to the form of eqn (3.38) as required.

At very low frequencies, waves on the ion cyclotron branch behave as follows. When $\omega \ll \omega_{ci}$, necessarily $\omega \ll \omega_{ce}$, and neglecting ω_{pi}^2 compared to ω_{pe}^2, eqn (3.78) gives

$$k = \frac{\omega}{c}\left(1 + \frac{\omega_{pe}^2}{\omega_{ce}\omega_{ci}}\right)^{\frac{1}{2}}. \tag{3.79}$$

Using eqn (3.72), this becomes

$$k = \frac{\omega}{c}\left(1 + \frac{\omega_{pi}^2}{\omega_{ci}^2}\right)^{\frac{1}{2}}. \tag{3.80}$$

Substituting for ω_{ci} and ω_{pi} from eqns (3.46) and (3.70) respectively, and using eqn (3.69), eqn (3.80) becomes

$$k = \frac{\omega}{c}\left\{1 + \frac{n_i M c^2}{(B^2/\mu_0)}\right\}^{\frac{1}{2}}. \tag{3.81}$$

The ratio of B^2/μ_0, which is twice the magnetic field energy density, to the mass density $n_i M$ of the plasma has the dimension of velocity squared. We therefore define the Alfvén velocity

$$V_A = (B^2/\mu_0 n_i M)^{\frac{1}{2}} \tag{3.82}$$

$$= (\omega_{ci}/\omega_{pi})c \tag{3.83}$$

$$= 6.9 \times 10^7 \left(\frac{B}{1\ \text{Tesla}}\right)\left(\frac{n_i}{10^{19}\ \text{m}^{-3}}\right)^{-\frac{1}{2}}\left(\frac{M}{m_p}\right)^{-\frac{1}{2}}\ \text{m s}^{-1}. \tag{3.84}$$

Combining eqns (3.81) and (3.82), the phase velocity of these low-frequency waves, which lie on the left circularly polarized ion cyclotron branch and propagate parallel to the magnetic field, is

$$\frac{\omega}{k} = \frac{V_A}{(1 + V_A^2/c^2)^{\frac{1}{2}}}. \tag{3.85}$$

In a typical plasma experiment, it follows from eqn (3.84) that $V_A^2/c^2 \ll 1$, so that eqn (3.85) gives $\omega/k \simeq V_A$. The Alfvén velocity is the sixth major plasma parameter that we have introduced. We have already seen that ω_{pe}, λ_D, and ω_{ce} characterize the high-frequency behaviour of plasmas, and that the ion frequencies ω_{pi} and ω_{ci} characterize the low-frequency behaviour. In the remainder of this section and throughout Chapter 4, we shall see repeated examples of the way in which V_A characterizes very-low-frequency, fluid-like plasma behaviour.

Next, we consider the right circularly polarized waves propagating parallel to the magnetic field, described by eqn (3.40). Substituting from eqns (3.73) and (3.74) for ε_1 and ε_2, we obtain

$$k = \frac{\omega}{c}\left\{1 + \frac{\omega_{pe}^2}{\omega(\omega_{ce} - \omega)} - \frac{\omega_{pi}^2}{\omega(\omega_{ci} + \omega)}\right\}^{\frac{1}{2}}. \tag{3.86}$$

Again using eqn (3.72), as we did after eqn (3.77), this becomes

$$k = \frac{\omega}{c}\left\{1 + \frac{\omega_{pe}^2 + \omega_{pi}^2}{(\omega_{ce} - \omega)(\omega_{ci} + \omega)}\right\}^{\frac{1}{2}}. \tag{3.87}$$

No new resonance appears, because the right circular polarization of the wave is opposite to the sense of gyration of the ions whose dynamics have now been included. However, eqn (3.87) enables us to extend our study of these modes to lower frequencies. First, let us neglect ω_{pi}^2 compared to ω_{pe}^2. Then when $\omega \gg \omega_{ci}$, the term $(\omega_{ci} + \omega)$ tends to ω, and Eq. (3.87) tends to eqn (3.41) as required. Let us now concentrate on the branch of solutions of eqn (3.87) with $\omega < \omega_{ce}$. When ω is large compared to ω_{ci}, this branch is referred to as the Whistler branch, described approximately by eqn (3.49). If $\omega \ll \omega_{ci}$, necessarily $\omega \ll \omega_{ce}$, and eqn (3.87) can be written approximately as

$$k = \frac{\omega}{c}\left(1 + \frac{\omega_{pi}^2}{\omega_{ce}\omega_{ci}}\right)^{\frac{1}{2}}. \tag{3.88}$$

This is identical to eqn (3.79), which was obtained for left circularly polarized waves. It follows that in this low-frequency limit, right circularly polarized waves also satisfy the dispersion relation eqn (3.85) that we derived from eqn (3.79). Thus, very low frequency waves on both the left circularly polarized ion cyclotron branch and the right circularly polarized Whistler branch have the same phase velocity V_A. They are both referred to as Alfvén waves: strictly speaking, they are* shear Alfvén waves, because the magnetic field perturbation due to the waves is always perpendicular to the background magnetic field. Conversely, any wave propagating along the magnetic field with phase velocity V_A may be resolved into left and right circularly polarized components.

Finally, let us turn to the low-frequency normal modes that propagate perpendicular to the magnetic field in a cold plasma. These are described by eqn (3.51), with ε_1, ε_2, and ε_3 given by eqns (3.73) to (3.75). The Ordinary mode dispersion relation eqn (3.52) is essentially unaltered by

* See Section 4.3.

the inclusion of ion dynamics; there merely appears an additional term ω_{pi}^2 in the right-hand side of eqn (3.53). The reasons for this lack of change are the same as those outlined in our discussion of eqn (3.76). Let us move on to the Extraordinary mode described by eqn (3.54). Because this equation is relatively complicated, we shall make approximations based on the relative magnitudes of different frequencies at an earlier stage than usual. We are concerned with low frequencies $\omega \ll \omega_{ce}$, so that to good approximation eqns (3.73) and (3.74) become

$$\varepsilon_1 = 1 + \frac{\omega_{pe}^2}{\omega_{ce}^2} + \frac{\omega_{pi}^2}{\omega_{ci}^2 - \omega^2}, \tag{3.89}$$

$$\varepsilon_2 = \frac{\omega_{pe}^2}{\omega_{ce}\omega} - \frac{\omega_{ci}}{\omega}\frac{\omega_{pi}^2}{\omega_{ci}^2 - \omega^2}. \tag{3.90}$$

Both of these expressions can be simplified using eqn (3.72). First, eqn (3.89) can be written

$$\varepsilon_1 = 1 + \frac{\omega_{pi}^2}{\omega_{ce}\omega_{ci}}\left(\frac{\omega_{ci}^2 + \omega_{ce}\omega_{ci} - \omega^2}{\omega_{ci}^2 - \omega^2}\right). \tag{3.91}$$

It is now convenient to define the lower hybrid frequency which, for our purposes, can be written*

$$\omega_{LH} = (\omega_{ce}\omega_{ci})^{\frac{1}{2}} \tag{3.92}$$

$$= 4.2 \times 10^9 \, Z^{\frac{1}{2}}\left(\frac{M}{m_p}\right)^{-\frac{1}{2}}\left(\frac{B}{1\ \text{Tesla}}\right) \text{rad s}^{-1}, \tag{3.93}$$

corresponding typically to frequencies of a few GHz. By eqn (3.71), ω_{LH} is comparable in magnitude to ω_{pi} in a typical experimental plasma. This is to be expected, since by eqns (3.72) and (3.92),

$$\omega_{pi} = (\omega_{pe}/\omega_{ce})\omega_{LH} \tag{3.94}$$

and in a typical experimental plasma ω_{pe}/ω_{ce} is of order unity. The lower hybrid frequency is, of course, lower than the upper hybrid frequency ω_{UH} defined by eqn (3.56). The hybrid aspect of ω_{LH} is the combination of electron and ion cyclotron motion, whereas in ω_{UH} it is the combination of electron space-charge and cyclotron motion. Clearly, from eqn (3.92), $\omega_{LH}^2 \gg \omega_{ci}^2$, and we now write eqn (3.91) in the form

$$\varepsilon_1 = 1 + \frac{\omega_{pe}^2}{\omega_{ce}^2}\frac{\omega_{LH}^2 - \omega^2}{\omega_{ci}^2 - \omega^2}. \tag{3.95}$$

Here we have used eqns (3.92) and (3.94). We can go further, and express ε_1 in terms of a common denominator. Using eqn (3.56), eqn

* This definition is valid when V_A^2/c^2 is much smaller than the ratio of electron mass to ion mass.

(3.95) becomes

$$\varepsilon_1 = \frac{\omega_{\mathrm{UH}}^2}{\omega_{\mathrm{ce}}^2(\omega_{\mathrm{ci}}^2 - \omega^2)} \left\{ \left(\frac{\omega_{\mathrm{pe}}^2 + \omega_{\mathrm{LH}}^2}{\omega_{\mathrm{pe}}^2 + \omega_{\mathrm{ce}}^2} \right) \omega_{\mathrm{LH}}^2 - \omega^2 \right\}. \tag{3.96}$$

It is convenient to define a critical frequency

$$\omega_{\mathrm{a}} = \{(\omega_{\mathrm{pe}}^2 + \omega_{\mathrm{LH}}^2)/(\omega_{\mathrm{pe}}^2 + \omega_{\mathrm{ce}}^2)\}^{\frac{1}{2}} \omega_{\mathrm{LH}}, \tag{3.97}$$

which is typically close to and smaller than ω_{LH}. Then ε_1 can finally be written

$$\varepsilon_1 = \frac{\omega_{\mathrm{UH}}^2}{\omega_{\mathrm{ce}}^2} \frac{(\omega_{\mathrm{a}}^2 - \omega^2)}{(\omega_{\mathrm{ci}}^2 - \omega^2)} \tag{3.98}$$

when $\omega \ll \omega_{\mathrm{ce}}$. The transformation of eqn (3.90) using eqn (3.72) is much simpler, and we obtain the following expression for ε_2 when $\omega \ll \omega_{\mathrm{ce}}$:

$$\varepsilon_2 = -\frac{\omega_{\mathrm{pi}}^2 \omega}{\omega_{\mathrm{ci}}(\omega_{\mathrm{ci}}^2 - \omega^2)}. \tag{3.99}$$

We now aim to substitute eqns (3.98) and (3.99) into eqn (3.54), to find the behaviour of the Extraordinary mode at frequencies well below ω_1, which is the low-frequency cut-off of the lower-frequency branch of the Extraordinary mode discussed in Section 3.3. Now eqn (3.54) gives

$$N^2 = (\varepsilon_1^2 - \varepsilon_2^2)/\varepsilon_1. \tag{3.100}$$

The existence of a real root N therefore depends on the magnitude of ε_1^2 compared to ε_2^2, and the sign of ε_1. By eqn (3.98), when ω is close to but smaller than ω_{a}, ε_1 is small and negative. At the same value of ω, by eqn (3.99), $\varepsilon_2 \simeq \omega_{\mathrm{pi}}^2/\omega_{\mathrm{ci}}\omega$ is positive and larger than unity. It follows that the right-hand side of eqn (3.100) is positive, and there is a real value of k given approximately by

$$k = \frac{\omega}{c} \left[\frac{\varepsilon_2}{|\varepsilon_1|} \right]^{\frac{1}{2}}. \tag{3.101}$$

From the expression eqn (3.98) for ε_1, we see that this third branch of Extraordinary mode has a resonance at its upper frequency limit, that is at $\omega_{\mathrm{a}} = \omega$ which is close to the lower hybrid frequency. This new branch, which owes its existence to ion dynamics, continues down to $\omega = 0$. When $\omega \ll \omega_{\mathrm{ci}}$, it follows from eqn (3.99) that ε_2 tends to zero. Then eqn (3.100) reduces to $N^2 = \varepsilon_1$. In this limit, ε_1 tends to $1 + \omega_{\mathrm{pi}}^2/\omega_{\mathrm{ci}}^2$, since the remaining contribution in eqn (3.89) is $\omega_{\mathrm{pe}}^2/\omega_{\mathrm{ce}}^2$, which is much smaller. Thus we have

$$k = \frac{\omega}{c} \left(1 + \frac{\omega_{\mathrm{pi}}^2}{\omega_{\mathrm{ci}}^2} \right)^{\frac{1}{2}}, \tag{3.102}$$

which is identical to eqn (3.80). Again, the dispersion relation eqn (3.81), or equivalently eqn (3.85), applies. In the low-frequency limit, the third branch of the Extraordinary mode, propagating perpendicular to the magnetic field, has a phase velocity close to V_A. In this respect, it resembles the low-frequency limit of the left and right circularly polarized waves (shear Alfvén waves) propagating parallel to the magnetic field. However, the perturbation of the magnetic field associated with the Extraordinary mode is directed along the magnetic field direction; this follows both from eqn (3.19) and from consideration of $\boldsymbol{M} \cdot \boldsymbol{E} = 0$. Thus, this wave perturbs the magnetic field strength, whereas shear Alfvén waves perturb the field direction. All the low-frequency plasma waves that extend down to $\omega \ll \omega_{ci}$ are shown in Fig. 3.4.

To conclude, we have now studied the normal modes of a cold magnetized plasma over the entire range of frequencies, from $\omega \gg \omega_{pe}$ to $\omega \ll \omega_{ci}$. Propagation both perpendicular and parallel to the magnetic field has been considered. For these two cases, the resolution of normal modes is conceptually and mathematically simple. At arbitrary angles of propagation, the position is more complicated, and the wave field is a combination of both classes of mode. Terms such as $N_x N_z$ in eqn (3.31), which are no longer zero, produce additional terms in the determinant which gives rise to the dispersion relation. However, further consideration of this topic is beyond the scope of this text. In the next chapter, we

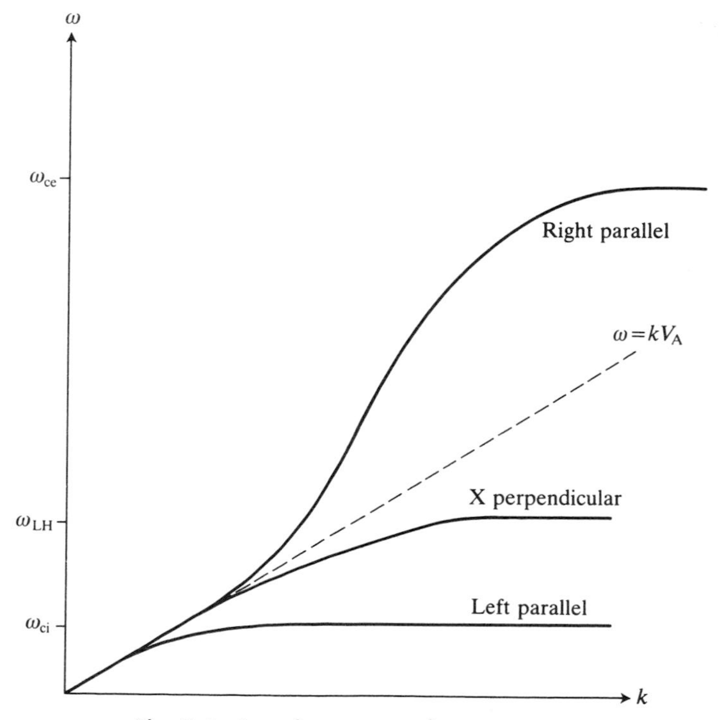

Fig. 3.4 Low-frequency plasma waves.

shall extend our study of low-frequency plasma behaviour by using a fluid rather than a dielectric model. This will give a different physical picture of the low frequency wave modes, and of the role of the Alfvén velocity.

Exercises

3.1. Consider a right circularly polarized wave with electric field amplitude $\sqrt{2}\,E_R$, and a left circularly polarized wave with amplitude $\sqrt{2}\,E_L$, which have the same frequency ω.

(a) If both waves travel through a plasma along a uniform magnetic field which is oriented in the z-direction, show that the total electric field is

$$\boldsymbol{E} = [\hat{\boldsymbol{e}}_x\{E_R \exp(ik_R z) + E_L \exp(ik_L z)\} + i\hat{\boldsymbol{e}}_y\{E_R \exp(ik_R z) - E_L \exp(ik_L z)\}]\exp(-i\omega t),$$

where k_R and k_L are given by eqns (3.41) and (3.38) respectively.

(b) Show that the ratio of the x and y components of the field, as a function of position z, is

$$\frac{E_x}{E_y} = -i\frac{[1 + (E_L/E_R)\exp\{i(k_L - k_R)z\}]}{[1 - (E_L/E_R)\exp\{i(k_L - k_R)z\}]}.$$

(c) A linearly polarized wave can be written as the superposition of two oppositely circularly polarized waves that have equal amplitudes. Show that, for the case considered, the plane of linear polarization rotates as the wave travels through the plasma (Faraday rotation). Show that the plane of linear polarization has rotated through a right-angle once the wave has travelled a distance $L = \pi/(k_L - k_R)$.

(d) For a high-frequency wave that has $\omega \gg \omega_{pe}$ and ω_{ce}, show that the angle of Faraday rotation is proportional to the product of electron density and magnetic field strength.

3.2. Wave energy is transmitted at the group velocity,

$$v_g = \partial\omega/\partial k.$$

(a) Obtain the group velocity for Whistler waves in the regime $\omega_{ce} \gg \omega$ and $\omega_{pe}^2/\omega_{ce}\omega \gg 1$.

(b) Show that if a packet of Whistler waves with a spread in frequency is created at a given instant, a distant observer will receive the high-frequency components earlier than the low-frequency components.

Solutions are on pages 149 *and* 150

4

Magnetohydrodynamic description of plasma

4.1 Introduction to magnetohydrodynamics

So far, we have based our study of plasmas on the single particle dynamics of the constituent electrons and ions. We have used these dynamics as the basis for a dielectric description of the plasma. This approach has been very effective. The understanding of any physical system is aided by knowledge of its normal modes, and the dielectric approach has enabled us to formulate the normal modes of plasmas over a very wide range of frequencies. In Section 3.4, we studied normal modes with phase velocity V_A defined by eqn (3.82), whose frequencies ω lie well below the lowest characteristic single particle frequency, the ion cyclotron frequency ω_{ci}. Two features of these modes suggest that an alternative approach to low frequency plasma behaviour might be fruitful. The first is the fact that V_A is determined by two charge-independent, macroscopic quantities: the mass density of the plasma, and the energy density of the magnetic field. Note that the mass density of the plasma is physically meaningful only when the plasma is considered on large lengthscales that embrace the positions of many ions. Averaged over these lengthscales, the plasma is electrically neutral. The second feature is that when $\omega \ll \omega_{ci}$, the timescales in question are much longer than those of any ion or electron oscillation associated with cyclotron or space charge motion. On these timescales, only the averaged, guiding-centre positions of the plasma particles are significant. It is clear that we have entered a regime of frequencies and lengthscales where a macroscopic, fluid-like description of the plasma would be more appropriate.

The question now arises, what sort of fluid model should we develop? What features can we distinguish, that will make the model apply specifically to plasmas? One set of answers is provided by the two-fluid description of plasmas in Chapter 6, where the plasma is taken as composed of two distinct but intermingled fluids, the electron and ion fluids, coupled together by their opposing electrical charges. Here, we shall adopt a simpler approach to the fact that at a microscopic level the plasma is composed of oppositely charged particles of unequal mass, but that at a macroscopic level it is electrically neutral. The plasma will be treated as a single, electrically neutral and perfectly conducting fluid. This reflects the fact that, on the timescales of interest, the high mobility of the electrons enables them both to keep the plasma electrically neutral and to cancel out any applied electric field. We shall need to distinguish between the macroscopic consequences of electron motion alone, which will support a current with volume density $\boldsymbol{J}$, and the macroscopic motion of the plasma as a whole, characterized by the bulk velocity $\boldsymbol{v}$. Note that $\boldsymbol{J}$ and $\boldsymbol{v}$ are independent. It is possible to visualize small elements of fluid moving in a given direction with velocity $\boldsymbol{v}$, while containing a current density $\boldsymbol{J}$ that is oriented in some other direction. The mass density of the fluid is denoted by ρ. At the microscopic level, ρ is given by the product of the ion number density n_i and ion mass M—since M greatly exceeds the electron mass m—but this information is not necessary in our present macroscopic approach. We are also in a position to make an advance on one aspect of the dielectric description of plasma in Chapter 3, where we restricted attention to cold plasmas. (We shall include thermal effects in Chapter 5.) Here, we shall allow the single fluid representing the plasma to possess a pressure p, which implies non-zero temperature.

Let us now set up the equations of ideal magnetohydrodynamics. First, we have the continuity equation

$$\frac{\partial \rho}{\partial t} + \nabla \cdot (\rho \boldsymbol{v}) = 0. \tag{4.1}$$

This expresses the conservation of mass, and is formally equivalent to eqn (I.14), which expresses the conservation of electric charge. Consider some closed volume V of space, and denote its surface by S. The total mass of the fluid contained in this volume is

$$M_f = \int_V \rho \, d^3\boldsymbol{x}. \tag{4.2}$$

Integrating eqn (4.1) over V, and applying the divergence theorem eqn (I.2) to the second term, we obtain an expression equivalent to eqn (I.15):

$$\frac{dM_f}{dt} = -\int_S \rho \boldsymbol{v} \cdot d\boldsymbol{S}. \tag{4.3}$$

This states that the rate of change with time of the mass of fluid contained within V is determined by the integral over the surface S of the normal component of the mass flux $\rho\boldsymbol{v}$; see Fig. 4.1. Note that $\rho\boldsymbol{v}$ has the dimensions of mass per unit time per unit area, or equivalently of momentum per unit volume.

Next, we require an equation for the time evolution of the fluid velocity $\boldsymbol{v}$. Both sides of this equation need careful consideration. We

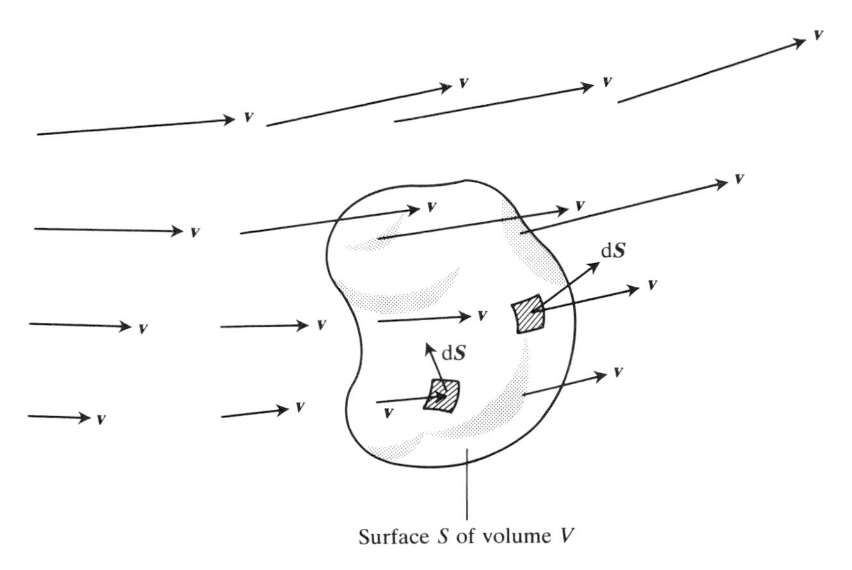

Fig. 4.1 Fluid flow through volume V.

wish to calculate the rate of change of the velocity of a particular fluid element in response to the forces acting on it. It will therefore be necessary to take a time derivative following the motion of the fluid element, just as we took a time derivative following the motion of a particular particle in eqns (2.26) to (2.29). Consider any function g which depends both on spatial position $\boldsymbol{x}$ and on time t. In the brief time interval Δt, a fluid element starting at $\boldsymbol{x}$ will move to a new position at $\boldsymbol{x} + \Delta t\boldsymbol{v}(\boldsymbol{x})$. The value of g appropriate to the new coordinates of the fluid element is, by Taylor's theorem,

$$g(\boldsymbol{x} + \Delta t\boldsymbol{v}(\boldsymbol{x}), t + \Delta t) \simeq g(\boldsymbol{x}, t) + \Delta t\boldsymbol{v}(\boldsymbol{x}) \cdot \boldsymbol{\nabla} g(\boldsymbol{x}, t) + \Delta t \frac{\partial}{\partial t} g(\boldsymbol{x}, t). \quad (4.4)$$

In eqn (4.4), the operator that is multiplied by Δt is the total or convective time derivative, following the motion of the fluid element, that we are seeking:

$$\frac{\mathrm{d}}{\mathrm{d}t} \equiv \frac{\partial}{\partial t} + \boldsymbol{v} \cdot \boldsymbol{\nabla}. \quad (4.5)$$

Thus, in our equation for the time evolution of the fluid velocity, the left-hand side will be given by $\mathrm{d}\boldsymbol{v}/\mathrm{d}t$, where the operator $\mathrm{d}/\mathrm{d}t$ is defined by eqn (4.5). The right hand side is determined by the forces acting on the fluid element. First, the fluid element contains a current density $\boldsymbol{J}$, so that in a magnetic field it will be subject to the force $\boldsymbol{J} \times \boldsymbol{B}/\rho$ per unit mass. Note that the fluid element is, by hypothesis, large enough to be electrically neutral, so that there is no Lorentz force associated with its velocity $\boldsymbol{v}$. Second, there is a pressure force $-(1/\rho)\nabla p$ per unit mass. Hence

$$\rho \frac{\mathrm{d}\boldsymbol{v}}{\mathrm{d}t} = \boldsymbol{J} \times \boldsymbol{B} - \nabla p. \tag{4.6}$$

Also, p and ρ are not independent. They are related by the standard equation of state, which for the behaviour that we shall be considering is $pV^{\gamma} = \text{constant}$, or

$$\frac{\mathrm{d}}{\mathrm{d}t}(p/\rho^{\gamma}) = 0, \tag{4.7}$$

where γ is the adiabatic exponent.

The three equations that we have considered so far, eqns (4.1), (4.6), and (4.7), govern the time evolution of the fluid quantities ρ, $\boldsymbol{v}$, and p, given a current $\boldsymbol{J}$ and magnetic field $\boldsymbol{B}$. In addition, we require complementary information: how $\boldsymbol{J}$ and $\boldsymbol{B}$ evolve in a given fluid. This is obtained as follows from Maxwell's equations and Ohm's law. Consider first Ohm's law for an element of a fluid that has conductivity σ. We have

$$\boldsymbol{J} = \sigma \boldsymbol{E}', \tag{4.8}$$

where $\boldsymbol{E}'$ is the electric field experienced by the fluid element in its rest frame. Now the fluid element is moving at velocity $\boldsymbol{v}$ with respect to the laboratory magnetic field $\boldsymbol{B}$. This motion gives rise to an electric field in the rest frame of the fluid element, which is described by the Lorentz transformation

$$\boldsymbol{E}' = \boldsymbol{E} + \boldsymbol{v} \times \boldsymbol{B}. \tag{4.9}$$

Here $\boldsymbol{E}$ denotes the electric field in the laboratory frame, and we have neglected all higher-order relativistic corrections because $v^2 \ll c^2$. Combining eqns (4.8) and (4.9), we have

$$\frac{1}{\sigma}\boldsymbol{J} = \boldsymbol{E} + \boldsymbol{v} \times \boldsymbol{B}. \tag{4.10}$$

In a perfectly conducting fluid, σ tends to infinity. This limiting case defines the subject area known as ideal magnetohydrodynamics, with which we shall be concerned in the rest of this chapter.* Then the left-hand side of eqn (4.10) tends to zero, and we have

$$\boldsymbol{E} = -\boldsymbol{v} \times \boldsymbol{B}. \tag{4.11}$$

Physically, eqn (4.11) can be considered in several ways. Combined with eqn (4.9), it tells us that the electric field $\boldsymbol{E}'$ in the rest frame of a perfectly conducting fluid is zero: the instantaneous response of the perfectly conducting fluid immediately cancels out any attempt to create a non-zero $\boldsymbol{E}'$. Now the laboratory is moving at velocity $-\boldsymbol{v}$ with respect to the rest frame of the fluid element, where the magnetic field $\boldsymbol{B}' = \boldsymbol{B}$,

* See, however, Exercise 4.1.

when terms of order v^2/c^2 are neglected. Then the Lorentz transformation back to the laboratory frame gives

$$\boldsymbol{E} = \boldsymbol{E}' - \boldsymbol{v} \times \boldsymbol{B}' = -\boldsymbol{v} \times \boldsymbol{B}, \tag{4.12}$$

since $\boldsymbol{E}' = 0$. Equation (4.12) is identical to eqn (4.11). Thus the existence of non-zero $\boldsymbol{E}$ has two physical origins. First, there is the assumed infinite conductivity of the fluid. Second, there is the fact that the division of an electromagnetic field into electric and magnetic components is frame-dependent. $\boldsymbol{E}'$ is zero, and $\boldsymbol{v} \times \boldsymbol{B}$ is non-zero, and this is sufficient to make $\boldsymbol{E}$ non-zero. In this sense, eqn (4.11) is a source equation for $\boldsymbol{E}$ in terms of $\boldsymbol{v}$ and $\boldsymbol{B}$.

Let us now include Faraday's law of electromagnetic induction eqn (I.4); $\nabla \times \boldsymbol{E} = -\partial \boldsymbol{B}/\partial t$. Taking the curl of eqn (4.11), this gives

$$\frac{\partial \boldsymbol{B}}{\partial t} = \nabla \times (\boldsymbol{v} \times \boldsymbol{B}). \tag{4.13}$$

We have now eliminated $\boldsymbol{E}$, and obtained in eqn (4.13) an expression for the time evolution of $\boldsymbol{B}$ in terms of $\boldsymbol{v}$ and $\boldsymbol{B}$ itself. However, we should remember that $\boldsymbol{E}$ plays a central role. The movement of the perfectly conducting fluid with respect to the magnetic field generates $\boldsymbol{E}$, and the curl of $\boldsymbol{E}$ in turn generates the time evolution of $\boldsymbol{B}$. Equation (4.13) is a combined statement of Ohm's law and Faraday's law for a perfectly conducting fluid in a magnetic field. It leads to the concept of magnetic flux freezing, which we now discuss. Consider a planar assembly of adjacent fluid elements which together form a surface S_1 at time t. At time $t + \Delta t$, these fluid elements have moved by amounts $\Delta t \boldsymbol{v}(\boldsymbol{x})$ to form a new surface S_2; see Fig. 4.2. Our aim is to establish a relation between the magnetic flux

$$\Phi_1 = \int_{S_1} \boldsymbol{B}(\boldsymbol{x}, t) \cdot \mathrm{d}\boldsymbol{S}_1 \tag{4.14}$$

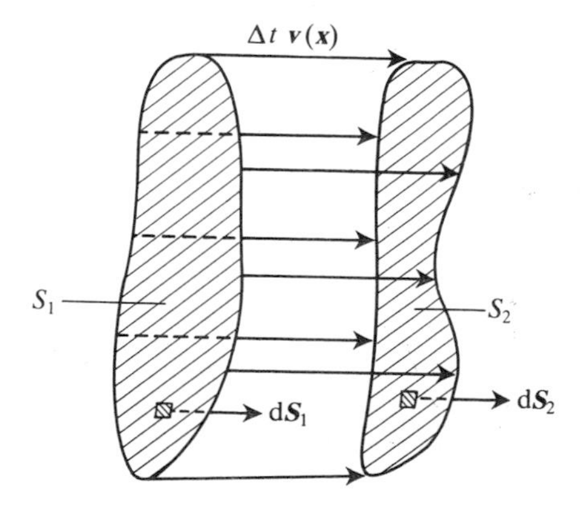

Fig. 4.2 New surface S_2 generated by fluid motion from S_1.

through the fluid surface S_1 and the magnetic flux

$$\Phi_2 = \int_{S_2} \boldsymbol{B}(\boldsymbol{x} + \Delta t \boldsymbol{v}(\boldsymbol{x}), t + \Delta t) \cdot \mathrm{d}\boldsymbol{S}_2 \tag{4.15}$$

through the surface S_2 defined by the positions of the same fluid elements at the later time. A first-order Taylor expansion of the explicitly time-dependent part of the integrand in eqn (4.15) gives

$$\Phi_2 = \int_{S_2} \boldsymbol{B}(\boldsymbol{x} + \Delta t \boldsymbol{v}(\boldsymbol{x}), t) \cdot \mathrm{d}\boldsymbol{S}_2 + \Delta t \int_{S_2} \frac{\partial}{\partial t} \boldsymbol{B}(\boldsymbol{x} + \Delta t \boldsymbol{v}(\boldsymbol{x}), t) \cdot \mathrm{d}\boldsymbol{S}_2. \tag{4.16}$$

As the second term in eqn (4.16) is already of order Δt, we neglect any further higher-order terms arising there, and obtain

$$\Phi_2 = \int_{S_2} \boldsymbol{B}(\boldsymbol{x} + \Delta t \boldsymbol{v}(\boldsymbol{x}), t) \cdot \mathrm{d}\boldsymbol{S}_2 + \Delta t \int_{S_1} \frac{\partial}{\partial t} \boldsymbol{B}(\boldsymbol{x}, t) \cdot \mathrm{d}\boldsymbol{S}_1. \tag{4.17}$$

We now recall one of the general properties of the magnetic field. It must satisfy $\boldsymbol{\nabla} \cdot \boldsymbol{B} = 0$, so that by the divergence theorem eqn (I.2), we have

$$\int_{\text{closed surface}} \boldsymbol{B} \cdot \mathrm{d}\boldsymbol{S} = 0. \tag{4.18}$$

Let us consider the particular closed surface at time t that is illustrated in Fig. 4.3. One wall is provided by S_1. A second wall is provided by S_2, the surface to which the fluid elements at S_1 will later move, displaced from S_1 by $\Delta t \boldsymbol{v}(\boldsymbol{x})$. The third surface S_3 is formed by the paths that will be taken by the fluid elements that form the boundary of S_1 to the boundary of S_2. By eqn (4.18), the integral at time t of $\boldsymbol{B} \cdot \mathrm{d}\boldsymbol{S}$ over the entire closed surface $\{S_1, S_2, S_3\}$ is zero. Recall that in eqn (4.18), $\mathrm{d}\boldsymbol{S}$ denotes the

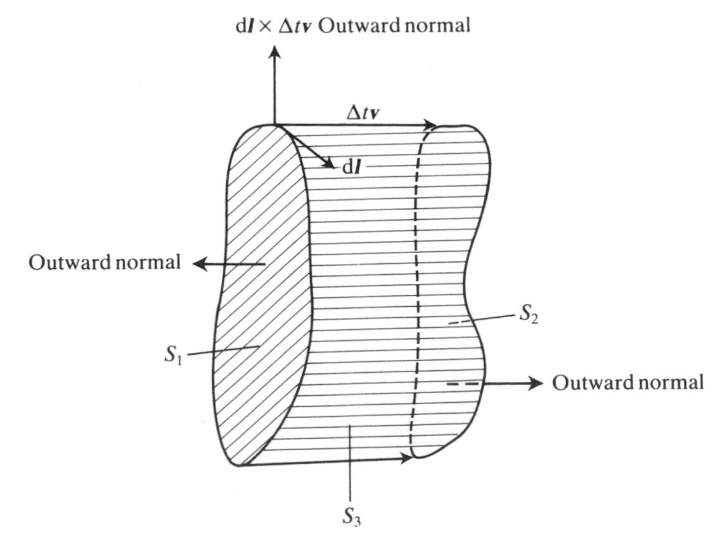

Fig. 4.3 Closed surface formed by S_1, S_2, and S_3.

outward normal. Comparing Figs. 4.2 and 4.3, we see that $\mathrm{d}\boldsymbol{S}_1$ in eqn (4.14) is an inward normal for the purposes of eqn (4.18). Then, for the closed surface of interest, we have from eqn (4.18)

$$\int_{S_1} \boldsymbol{B}(\boldsymbol{x}, t) \cdot \mathrm{d}\boldsymbol{S}_1 = \int_{S_2} \boldsymbol{B}(\boldsymbol{x} + \Delta t \boldsymbol{v}(\boldsymbol{x}), t) \cdot \mathrm{d}\boldsymbol{S}_2 + \int_{S_3} \boldsymbol{B}(\boldsymbol{x}, t) \cdot \mathrm{d}\boldsymbol{S}_3. \quad (4.19)$$

Here we have taken the value of $\boldsymbol{B}$ appropriate to each surface. Equation (4.19) is simply a consequence of $\nabla \cdot \boldsymbol{B} = 0$. Now eqn (4.14) enables us to replace the left-hand side of eqn (4.19) by Φ_1, and we can then use eqn (4.19) to eliminate the term $\int_{S_2} \boldsymbol{B}(\boldsymbol{x} + \Delta t \boldsymbol{v}(x), t) \cdot \mathrm{d}\boldsymbol{S}_2$ in eqn (4.17). This gives

$$\Phi_2 = \Phi_1 + \Delta t \int_{S_1} \frac{\partial}{\partial t} \boldsymbol{B}(\boldsymbol{x}, t) \cdot \mathrm{d}\boldsymbol{S}_1 - \int_{S_3} \boldsymbol{B}(\boldsymbol{x}, t) \cdot \mathrm{d}\boldsymbol{S}_3. \quad (4.20)$$

It can be seen from Fig. 4.3 that $\mathrm{d}\boldsymbol{S}_3 = \mathrm{d}\boldsymbol{l} \times \Delta t \boldsymbol{v}(\boldsymbol{x})$, so that

$$\begin{aligned} \int_{S_3} \boldsymbol{B}(\boldsymbol{x}, t) \cdot \mathrm{d}\boldsymbol{S}_3 &= \int \boldsymbol{B}(\boldsymbol{x}, t) \cdot (\mathrm{d}\boldsymbol{l} \times \Delta t \boldsymbol{v}(\boldsymbol{x})) \\ &= \Delta t \oint_C \{\boldsymbol{v}(\boldsymbol{x}) \times \boldsymbol{B}(\boldsymbol{x}, t)\} \cdot \mathrm{d}\boldsymbol{l}, \end{aligned} \quad (4.21)$$

where the contour C is the common boundary of S_1 and S_3, and we have used the vector identity $\boldsymbol{a} \cdot (\boldsymbol{b} \times \boldsymbol{c}) = \boldsymbol{b} \cdot (\boldsymbol{c} \times \boldsymbol{a})$. Next we use Stokes' theorem eqn (I.5) on the right-hand side of eqn (4.21):

$$\int_{S_3} \boldsymbol{B}(\boldsymbol{x}, t) \cdot \mathrm{d}\boldsymbol{S}_3 = \Delta t \int_{S_1} \nabla \times (\boldsymbol{v} \times \boldsymbol{B}) \cdot \mathrm{d}\boldsymbol{S}_1, \quad (4.22)$$

where, for brevity, we have ceased to write arguments $(\boldsymbol{x}, t)$ explicitly. Finally, substituting eqn (4.22) into eqn (4.20), we obtain

$$\Phi_2 - \Phi_1 = \Delta t \int_{S_1} \left\{ \frac{\partial \boldsymbol{B}}{\partial t} - \nabla \times (\boldsymbol{v} \times \boldsymbol{B}) \right\} \cdot \mathrm{d}\boldsymbol{S}_1. \quad (4.23)$$

By eqn (4.13), the integrand in eqn (4.23) vanishes everywhere, so that

$$\Phi_2 = \Phi_1. \quad (4.24)$$

This states the principle of magnetic flux freezing in a perfectly conducting fluid: the magnetic flux through a surface defined by a set of adjacent fluid elements does not change as the surface moves and changes shape following the flow $\boldsymbol{v}(\boldsymbol{x}, t)$ of the fluid elements. This is a consequence of Ohm's law for a perfectly conducting fluid, together with two of Maxwell's equations, eqns (I.4) and (I.7): $\nabla \times \boldsymbol{E} = -\partial \boldsymbol{B}/\partial t$ and $\nabla \cdot \boldsymbol{B} = 0$.

The final equation of ideal magnetohydrodynamics relates the spatial distribution of the magnetic field to its source. Recall Maxwell's equation eqn (I.9), $\nabla \times \boldsymbol{H} = \boldsymbol{J} + \varepsilon_0 \partial \boldsymbol{E}/\partial t$. We first show that at the non-relativistic velocities $\boldsymbol{v}$ that we consider, the contribution from the displacement current is negligible. Using eqn (4.11) for $\boldsymbol{E}$, we have $\varepsilon_0 \partial \boldsymbol{E}/\partial t = -(1/\mu_0 c^2)\partial/\partial t(\boldsymbol{v} \times \boldsymbol{B})$. Consider, for example, the contribution

$$\frac{-1}{\mu_0 c^2}\boldsymbol{v} \times \frac{\partial \boldsymbol{B}}{\partial t} = \frac{-1}{c^2}\boldsymbol{v} \times \nabla \times (\boldsymbol{v} \times \boldsymbol{H}). \tag{4.25}$$

This is smaller than the $\nabla \times \boldsymbol{H}$ term on the left-hand side of Maxwell's equation eqn (I.9) by a factor of order v^2/c^2. To be fully justified in neglecting the displacement current, we also require $V_A/c \ll 1$, where the Alfvén velocity V_A was defined in eqn (3.82). Anticipating the results of the next two sections, suppose that there exist magnetohydrodynamic oscillations whose phase velocity is close to the Alfvén velocity—we have met such oscillations before in eqn (3.85). Then the oscillating part of the magnetic field (as opposed to the background equilibrium component) satisfies $|\partial \boldsymbol{B}/\partial t| \sim \omega\,|\boldsymbol{B}| \sim kV_A\,|\boldsymbol{B}| \sim V_A\,|\nabla \times \boldsymbol{B}|$; using eqn (4.25) and the preceding discussion, it then follows that $\varepsilon_0\,|\partial \boldsymbol{E}/\partial t| \sim (V_A v/c^2)\,|\nabla \times \boldsymbol{H}|$. Thus, so long as v^2/c^2 and V_A/c are both much smaller than unity, we may use Ampère's law for the magnetic source equation:

$$\nabla \times \boldsymbol{B} = \mu_0 \boldsymbol{J}. \tag{4.26}$$

This completes the set of equations of ideal magnetohydrodynamics. The coupled evolution of the variables $\boldsymbol{v}$, $\boldsymbol{J}$, ρ, p, and $\boldsymbol{B}$ is governed by eqns (4.1), (4.6), (4.7), (4.13), and (4.26).

4.2 Force and motion in ideal magnetohydrodynamics

Our first step in investigating the equations of ideal magnetohydrodynamics is to combine them in a way which reduces the number of independent variables. Using the definition eqn (4.5) of the convective time derivative, the equation of continuity eqn (4.1) can be written

$$\frac{\mathrm{d}\rho}{\mathrm{d}t} = -\rho \nabla \cdot \boldsymbol{v}. \tag{4.27}$$

For the particular case of an incompressible fluid, where ρ cannot change, the vanishing of the left-hand side of eqn (4.27) implies

$$\nabla \cdot \boldsymbol{v} = 0. \tag{4.28}$$

The equation of state eqn (4.7) can be written

$$\frac{\mathrm{d}p}{\mathrm{d}t} = \frac{\gamma p}{\rho}\frac{\mathrm{d}\rho}{\mathrm{d}t}. \tag{4.29}$$

We can now eliminate $\mathrm{d}\rho/\mathrm{d}t$ between eqns (4.27) and (4.29). This gives

$$\frac{\mathrm{d}p}{\mathrm{d}t} = -\gamma p \nabla \cdot \boldsymbol{v}, \tag{4.30}$$

which expresses the time evolution of p in terms only of $\boldsymbol{v}$ and p itself.

We now turn to eqn (4.13), which is a combination of Ohm's law and Maxwell's equations. It can be recast using the identity

$$\nabla \times (\boldsymbol{A} \times \boldsymbol{B}) = \boldsymbol{A}(\nabla \cdot \boldsymbol{B}) - \boldsymbol{B}(\nabla \cdot \boldsymbol{A}) + (\boldsymbol{B} \cdot \nabla)\boldsymbol{A} - (\boldsymbol{A} \cdot \nabla)\boldsymbol{B}. \quad (4.31)$$

Again using eqn (4.5), this identity enables us to write eqn (4.13) in the form

$$\frac{\mathrm{d}\boldsymbol{B}}{\mathrm{d}t} = -\boldsymbol{B}(\nabla \cdot \boldsymbol{v}) + (\boldsymbol{B} \cdot \nabla)\boldsymbol{v}, \quad (4.32)$$

where we have used the fact that $\nabla \cdot \boldsymbol{B} = 0$. Note that in the particular case of an incompressible fluid, eqn (4.28) implies that the first term on the right-hand side of eqn (4.32) also vanishes, so that

$$\frac{\mathrm{d}\boldsymbol{B}}{\mathrm{d}t} = (\boldsymbol{B} \cdot \nabla)\boldsymbol{v}. \quad (4.33)$$

In the general case, where eqn (4.27) applies, eqn (4.32) gives

$$\frac{\mathrm{d}\boldsymbol{B}}{\mathrm{d}t} - \frac{\boldsymbol{B}}{\rho}\frac{\mathrm{d}\rho}{\mathrm{d}t} = (\boldsymbol{B} \cdot \nabla)\boldsymbol{v}. \quad (4.34)$$

Dividing eqn (4.34) by ρ, we obtain

$$\frac{\mathrm{d}}{\mathrm{d}t}\left(\frac{\boldsymbol{B}}{\rho}\right) = \left\{\left(\frac{\boldsymbol{B}}{\rho}\right) \cdot \nabla\right\}\boldsymbol{v}. \quad (4.35)$$

Note that eqns (4.33) and (4.35) have the same mathematical form, with the quantity $\boldsymbol{B}/\rho$ in the general case eqn (4.35) playing the same role as $\boldsymbol{B}$ in the incompressible case eqn (4.33). Both equations were obtained by combining the continuity equation with eqn (4.13), which gives rise to flux freezing. It is therefore of interest to establish the connection between the form of eqn (4.33), or equivalently eqn (4.35), and the general properties of the fluid motion. Let the vector $\boldsymbol{G}$ stand for $\boldsymbol{B}$ or $\boldsymbol{B}/\rho$ as appropriate, and let G_i be the ith component of $\boldsymbol{G}$; that is, i stands for x, y or z. Then the general form of eqns (4.33) and (4.35) is

$$\frac{\mathrm{d}G_i}{\mathrm{d}t} = G_j \frac{\partial v_i}{\partial x_j}, \quad (4.36)$$

where summation over the repeated index j is understood. Between times t and $t + \Delta t$, the fluid element whose coordinates are initially $x_i(t)$ moves to

$$x_i(t + \Delta t) = x_i(t) + \Delta t v_i(t). \quad (4.37)$$

Now $\mathrm{d}\boldsymbol{G}/\mathrm{d}t$ denotes the convective derivatives of $\boldsymbol{G}$ following the fluid motion. Therefore the value of each component G_i at the position occupied by the moving fluid element changes from $G_i(\boldsymbol{x}(t), t)$ to

$$G_i(\boldsymbol{x}(t + \Delta t), t + \Delta t) = G_i(\boldsymbol{x}(t), t) + \Delta t \frac{\mathrm{d}G_i}{\mathrm{d}t}. \quad (4.38)$$

Substituting for $\mathrm{d}G_i/\mathrm{d}t$ from eqn (4.36), eqn (4.38) becomes

$$G_i(\boldsymbol{x}(t+\Delta t), t+\Delta t) = G_j(\boldsymbol{x}(t), t)\left(\delta_{ij} + \Delta t \frac{\partial v_i(t)}{\partial x_j(t)}\right). \tag{4.39}$$

Here δ_{ij}, the Kronecker delta, takes the value one if $i=j$ and is zero otherwise. We note that the various components of the position vector $\boldsymbol{x}(t)$ are independent of each other, so that

$$\frac{\partial x_i(t)}{\partial x_j(t)} = \delta_{ij}. \tag{4.40}$$

We can use this expression to replace δ_{ij} in eqn (4.39), giving

$$G_i(\boldsymbol{x}(t+\Delta t), t+\Delta t) = G_j(\boldsymbol{x}(t), t)\left(\frac{\partial x_i(t)}{\partial x_j(t)} + \Delta t \frac{\partial v_i(t)}{\partial x_j(t)}\right). \tag{4.41}$$

Finally we use eqn (4.37) to write eqn (4.41) in the form

$$G_i(\boldsymbol{x}(t+\Delta t), t+\Delta t) = G_j(\boldsymbol{x}(t), t)\frac{\partial x_i(t+\Delta t)}{\partial x_j(t)}. \tag{4.42}$$

The tensor $\partial x_i(t+\Delta t)/\partial x_j(t)$ is the stress tensor of the instantaneous fluid motion, which characterizes the relative positions of the fluid element at the beginning and end of the time interval. Its occurrence in eqn (4.42) makes this a very powerful result. The time evolution of $\boldsymbol{G}$ following the fluid element is determined exclusively by the initial value of $\boldsymbol{G}$ and the local stress tensor of the fluid; the value of $\boldsymbol{G}$ in neighbouring fluid elements is not relevant. Since $\boldsymbol{G}$ denotes $\boldsymbol{B}$ or $\boldsymbol{B}/\rho$, this result is another expression of the concept of magnetic flux freezing.

Thus far, we have used three of our ideal magnetohydrodynamic equations, namely eqns (4.1), (4.7), and (4.13), to study magnetic field kinematics: the way in which a magnetic field moves in a perfectly conducting fluid that is also in motion. Both in this section and in the preceding one, we have seen how the concept of magnetic flux freezing arises naturally in this context. Let us now move on to consider the dynamics of the system: the forces that a magnetic field exerts on a fluid in which it is present. We take the equation of motion for the fluid, eqn (4.6), and express the current density $\boldsymbol{J}$ in terms of the curl of the magnetic field using eqn (4.26). Then we have

$$\rho\frac{\mathrm{d}\boldsymbol{v}}{\mathrm{d}t} = \frac{1}{\mu_0}(\nabla\times\boldsymbol{B})\times\boldsymbol{B} - \nabla p. \tag{4.43}$$

Using the identity

$$(\nabla\times\boldsymbol{B})\times\boldsymbol{A} = (\boldsymbol{A}\cdot\nabla)\boldsymbol{B} - (\nabla\boldsymbol{B})\cdot\boldsymbol{A}, \tag{4.44}$$

eqn (4.43) becomes

$$\rho\frac{\mathrm{d}\boldsymbol{v}}{\mathrm{d}t} = -\nabla\left(p + \frac{B^2}{2\mu_0}\right) + \frac{1}{\mu_0}(\boldsymbol{B}\cdot\nabla)\boldsymbol{B}. \tag{4.45}$$

This is a fundamental equation. It shows how the presence in a conducting fluid of a magnetic field with non-zero curl is in general sufficient to set the fluid in motion. Together with eqn (4.13), which shows how the motion of a conducting fluid in a magnetic field causes the field to evolve with time, eqn (4.45) indicates the way in which much of the large-scale non-equilibrium behaviour of plasmas arises.

The final term in eqn (4.45), $(1/\mu_0)(\boldsymbol{B}\cdot\boldsymbol{\nabla})\boldsymbol{B}$, is associated with the rate of change that is observed in the magnetic field $\boldsymbol{B}$ as we move along it. Let us use s as a distance parameter along the magnetic field, so that

$$(\boldsymbol{B}\cdot\boldsymbol{\nabla}) = B\frac{\partial}{\partial s}. \tag{4.46}$$

Now $\boldsymbol{B}$, on which $(\boldsymbol{B}\cdot\boldsymbol{\nabla})$ operates, may be changing both in strength and in direction. We write

$$\boldsymbol{B} = B\hat{\boldsymbol{b}} \tag{4.47}$$

where $\hat{\boldsymbol{b}}$ is a unit vector in the local direction of the magnetic field. Combining eqns (4.46) and (4.47), we have

$$\begin{aligned}\frac{1}{\mu_0}(\boldsymbol{B}\cdot\boldsymbol{\nabla})\boldsymbol{B} &= \frac{1}{\mu_0}\left\{\left(B\frac{\partial B}{\partial s}\right)\hat{\boldsymbol{b}} + B^2\frac{\partial\hat{\boldsymbol{b}}}{\partial s}\right\}\\ &= \frac{B^2}{\mu_0}\frac{\partial\hat{\boldsymbol{b}}}{\partial s} + \hat{\boldsymbol{b}}\frac{\partial}{\partial s}\left(\frac{B^2}{2\mu_0}\right).\end{aligned} \tag{4.48}$$

The first term on the right in eqn (4.48) depends on the change in direction of the magnetic field as we move along it. We define

$$\frac{\partial\hat{\boldsymbol{b}}}{\partial s} = \frac{\hat{\boldsymbol{n}}}{R} \tag{4.49}$$

where, as shown in Fig. (4.4), $\hat{\boldsymbol{n}}$ is the principal normal to the magnetic field line and R is its local radius of curvature. The second term in eqn (4.48) is equal in magnitude and opposite in sign to the component parallel to the magnetic field of the term $-\boldsymbol{\nabla}(B^2/2\mu_0)$ in eqn (4.45). It is therefore convenient to define a gradient operator which acts only perpendicular to the local magnetic field direction:

$$\boldsymbol{\nabla}_\perp = \boldsymbol{\nabla} - \hat{\boldsymbol{b}}\frac{\partial}{\partial s}. \tag{4.50}$$

Combining eqns (4.48) to (4.50), eqn (4.45) becomes

$$\rho\frac{\mathrm{d}\boldsymbol{v}}{\mathrm{d}t} = -\boldsymbol{\nabla}p - \boldsymbol{\nabla}_\perp\left(\frac{B^2}{2\mu_0}\right) + \frac{1}{R}\left(\frac{B^2}{\mu_0}\right)\hat{\boldsymbol{n}}. \tag{4.51}$$

We see from eqn (4.51) that the magnetic field exerts two types of force on the fluid. As we expect, neither of these forces has any component parallel to the magnetic field itself. First, there is the term $-\boldsymbol{\nabla}_\perp(B^2/2\mu_0)$. This is a perpendicular magnetic pressure term; it acts to drive the fluid away from regions of high magnetic field strength. Second, there is the

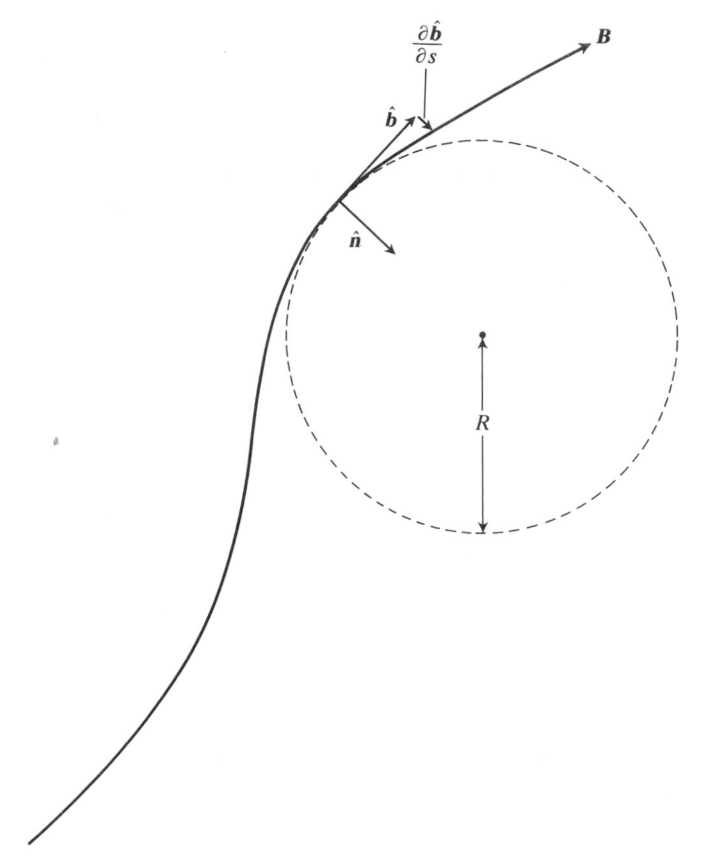

Fig. 4.4 $\partial \hat{\boldsymbol{b}}/\partial s$ in a magnetic field $\boldsymbol{B}$.

term $(1/R)(B^2/\mu_0)\hat{\boldsymbol{n}}$. The sign of this term is such that it accelerates the fluid along $\hat{\boldsymbol{n}}$ towards the centre of curvature of the field line; see again Fig. 4.4. It follows from the concept of magnetic flux freezing that the resulting fluid motion will pull the magnetic field line with it. Thus, this force acts to straighten the field line, reducing its curvature. In an analogous manner, the tension in a stretched string acts to straighten it when it is plucked, so that the final term in eqn (4.51) describes a tension in magnetic field lines. This tension is additional to the magnetic pressure described by the second term in eqn (4.51).

These considerations lead naturally to the topic of the types of wave motion that satisfy the equations of ideal magnetohydrodynamics. Before examining these oscillations in the next section, it is useful to consider some basic properties of the equations. For simplicity, we take an incompressible fluid for which eqn (4.28) applies. The fact that the fluid density ρ is a constant means that the force equation (4.45) can be written

$$\left(\frac{\partial}{\partial t}+\boldsymbol{v}\cdot\boldsymbol{\nabla}\right)\boldsymbol{v}=-\boldsymbol{\nabla}\left(\frac{p}{\rho}+\frac{B^2}{2\mu_0\rho}\right)+\frac{1}{\mu_0\rho}(\boldsymbol{B}\cdot\boldsymbol{\nabla})\boldsymbol{B}. \tag{4.52}$$

At this stage, a quantity familiar from our earlier treatment of plasma oscillations has reappeared. Referring back to eqn (3.82), we define the vector Alfvén velocity

$$\boldsymbol{V}_{\mathrm{A}}=\boldsymbol{B}/(\mu_0\rho)^{\frac{1}{2}}. \tag{4.53}$$

Then eqn (4.52) becomes

$$\left(\frac{\partial}{\partial t}+\boldsymbol{v}\cdot\boldsymbol{\nabla}\right)\boldsymbol{v}=-\boldsymbol{\nabla}\left(\frac{p}{\rho}+\tfrac{1}{2}V_{\mathrm{A}}^2\right)+(\boldsymbol{V}_{\mathrm{A}}\cdot\boldsymbol{\nabla})\boldsymbol{V}_{\mathrm{A}}. \tag{4.54}$$

Again using the constancy of ρ, we may write eqn (4.33), which describes the time evolution of the magnetic field in an incompressible perfectly conducting fluid, in the form

$$\left(\frac{\partial}{\partial t}+\boldsymbol{v}\cdot\boldsymbol{\nabla}\right)\boldsymbol{V}_{\mathrm{A}}=(\boldsymbol{V}_{\mathrm{A}}\cdot\boldsymbol{\nabla})\boldsymbol{v}. \tag{4.55}$$

Equations (4.54) and (4.55) are a pair of coupled evolution equations for the fluid velocity $\boldsymbol{v}$ and the local Alfvén velocity $\boldsymbol{V}_{\mathrm{A}}$. To emphasize their common features, let us write these equations in the form

$$\begin{bmatrix}\dfrac{\partial}{\partial t}+\boldsymbol{v}\cdot\boldsymbol{\nabla} & -\mathbf{V}_{\mathrm{A}}\cdot\boldsymbol{\nabla}\\[2ex] -\boldsymbol{V}_{\mathrm{A}}\cdot\boldsymbol{\nabla} & \dfrac{\partial}{\partial t}+\boldsymbol{v}\cdot\boldsymbol{\nabla}\end{bmatrix}\begin{bmatrix}\boldsymbol{v}\\[2ex] \boldsymbol{V}_{\mathrm{A}}\end{bmatrix}=\begin{bmatrix}-\dfrac{1}{\rho}\boldsymbol{\nabla}(p+\tfrac{1}{2}\rho V_{\mathrm{A}}^2)\\[2ex] 0\end{bmatrix}. \tag{4.56}$$

We shall now show that magnetohydrodynamic equilibrium is possible. That is, the equations of ideal magnetohydrodynamics, which combine to give eqn (4.56), have solutions for which $\partial/\partial t\to 0$. For example, suppose that $(\boldsymbol{B}\cdot\boldsymbol{\nabla})\,\boldsymbol{B}=0$ and

$$p+\tfrac{1}{2}\rho V_{\mathrm{A}}^2=\text{constant}=p+\frac{B^2}{2\mu_0}. \tag{4.57}$$

In this case, eqn (4.56) can be satisfied in two ways. We may have

$$\boldsymbol{v}=0 \tag{4.58}$$

everywhere, so that the fluid is motionless. Alternatively, we may have $\boldsymbol{v}=\pm\boldsymbol{V}_{\mathrm{A}}$, so that the fluid flows along the magnetic field at the local Alfvén velocity.

The equilibrium which consists of a static, motionless fluid, described by eqns (4.57) and (4.58), is clearly of great practical interest. Referring back to eqns (4.45) and (4.6), this equilibrium occurs when the distribution of current and magnetic field in the fluid sets up a $\boldsymbol{J} \times \boldsymbol{B}$ force that exactly balances the pressure gradient force at every point. This equilibrium need not, of course, be stable. Its stability is determined by the nature of the forces that arise when the fluid is slightly displaced. If these forces act to restore the equilibrium, it is stable. In this case, the fluid will respond to perturbation by oscillating. The characteristic normal modes of this oscillation will be discussed in the following section. On the other hand, it may be possible to reach a state of lower energy when the equilibrium configuration is given a small perturbation. In this case, the equilibrium is unstable. We shall discuss a general magnetohydrodynamic energy principle in Section 4.4.

4.3 Magnetohydrodynamic waves

There are three basic physical conditions required for wave motion in a continuous medium. These were exemplified in Section 1.1, where we introduced the electron plasma frequency. First, we require an equilibrium configuration, with respect to which small displacements can be considered. Next, we require restoring forces that act to reverse the displacement. Finally, it is necessary for the medium to possess inertia, so that the kinetic energy of the returning displaced fluid element carries it back past its equilibrium position, reversing its initial displacement. All three conditions are met by the equations of ideal magnetohydrodynamics. We saw at the conclusion of the preceding section that equilibria exist. Restoring forces can arise from the thermal pressure, magnetic pressure, and magnetic tension described by eqn (4.45), or equivalently eqn (4.51). Inertia is provided by the fluid mass density ρ.

The study of magnetohydrodynamic waves is an extension of the study of low-frequency plasma normal modes that we commenced in Section 3.4. There, we saw how a characteristic phase velocity, the Alfvén velocity V_A given by eqn (3.82), arises naturally from the electron and ion contributions to the plasma dielectric tensor in the low-frequency regime. This velocity reappeared in a different context at the end of Section 4.2, where the equations of incompressible ideal magnetohydrodynamics were formulated with the vector Alfvén velocity $\boldsymbol{V}_A = V_A \hat{\boldsymbol{b}}$, given by eqn (4.53), as a basic variable. There are accordingly good reasons, that have appeared before we have considered magnetohydrodynamic waves as such, to expect the Alfvén velocity to play a central role in this topic.

First, let us construct our equilibrium configuration, for which time derivatives vanish. Denoting equilibrium quantities by subscript zero, we shall assume that at equilibrium the plasma is at rest,

$$\boldsymbol{v}_0 = 0. \tag{4.59}$$

By eqn (4.45), this requires

$$\nabla\left(p_0 + \frac{B_0^2}{2\mu_0}\right) = 0, \tag{4.60}$$

$$(\boldsymbol{B}_0 \cdot \nabla)\boldsymbol{B}_0 = 0. \tag{4.61}$$

When substituted in eqns (4.27), (4.30), and (4.35) respectively, eqn (4.59) guarantees that ρ_0, p_0, and $\boldsymbol{B}_0$ are indeed independent of time. We now consider first-order perturbations of this equilibrium, writing fluid quantities in the form $\rho = \rho_0 + \delta\rho$. In the linearized regime where we neglect products of first-order perturbations on grounds of smallness, eqn (4.27) gives

$$\frac{\partial}{\partial t}\delta\rho = -\rho_0 \nabla \cdot \boldsymbol{v}. \tag{4.62}$$

Here we have written $\boldsymbol{v}$ for $\delta\boldsymbol{v}$ since $\boldsymbol{v}_0$ is zero. We now define the sound speed

$$V_s = (\gamma p_0/\rho_0)^{\frac{1}{2}} \tag{4.63}$$

$$= 5.4 \times 10^5 \left(\frac{\gamma}{3}\right)\left(\frac{k_B T}{1\ \text{keV}}\right)^{\frac{1}{2}}\left(\frac{M}{m_p}\right)^{-\frac{1}{2}} \text{m s}^{-1}. \tag{4.64}$$

We note that, for fusion plasmas and diffuse space plasmas, the magnitudes of V_A, given by eqn (3.84), and of V_s are typically within a few orders of magnitude of the velocity of light. Since $\boldsymbol{v}_0$ is zero, it is reasonable to consider a regime where

$$|\boldsymbol{v}| \ll V_A, V_s. \tag{4.65}$$

Using eqn (4.63), eqns (4.29) and (4.62) give

$$\frac{\partial}{\partial t}\delta p = V_s^2 \frac{\partial}{\partial t}\delta\rho = -\rho_0 V_s^2(\nabla \cdot \boldsymbol{v}). \tag{4.66}$$

Linearized first-order perturbations of the magnetic field and of the fluid velocity satisfy evolution equations that follow from eqns (4.32) and (4.45) respectively:

$$\frac{\partial}{\partial t}\delta\boldsymbol{B} = -\boldsymbol{B}_0(\nabla \cdot \boldsymbol{v}) + (\boldsymbol{B}_0 \cdot \nabla)\boldsymbol{v}, \tag{4.67}$$

$$\rho_0 \frac{\partial \boldsymbol{v}}{\partial t} = -\nabla\left(\delta p + \frac{\boldsymbol{B}_0 \cdot \delta\boldsymbol{B}}{\mu_0}\right) + \frac{1}{\mu_0}(\boldsymbol{B}_0 \cdot \nabla)\delta\boldsymbol{B}. \tag{4.68}$$

The final term in eqn (4.45), proportional to $(\boldsymbol{B} \cdot \nabla)\boldsymbol{B}$, also gives rise to a term $(\delta\mathbf{B} \cdot \nabla)\boldsymbol{B}_0$. We assume that gradients in $\boldsymbol{B}_0$ are sufficiently weak that this term can be neglected. That is, it is at least as small as the other terms that we have neglected, involving products of first order perturbations. For convenience, we shall define the z-axis to lie along the equilibrium magnetic field direction, so that

$$\boldsymbol{B}_0 = B_0\hat{\boldsymbol{e}}_z; \qquad (\boldsymbol{B}_0 \cdot \boldsymbol{\nabla}) = B_0\frac{\partial}{\partial z}; \qquad \boldsymbol{B}_0 \cdot \delta\boldsymbol{B} = B_0\delta B_z. \tag{4.69}$$

Then eqns (4.67) and (4.68) can be written

$$\frac{\partial}{\partial t}\delta\boldsymbol{B} = -B_0\left\{(\boldsymbol{\nabla}\cdot\boldsymbol{v})\hat{\boldsymbol{e}}_z - \frac{\partial \boldsymbol{v}}{\partial z}\right\} \tag{4.70}$$

$$\frac{\partial \boldsymbol{v}}{\partial t} = -\frac{1}{\rho_0}\boldsymbol{\nabla}\delta p - \frac{V_\mathrm{A}^2}{B_0}\left(\boldsymbol{\nabla}\delta B_z - \frac{\partial}{\partial z}\delta\boldsymbol{B}\right), \tag{4.71}$$

where we have used the fact that by eqn (4.53), $V_\mathrm{A}^2 = B_0^2/\mu_0\rho_0$.

We now have a closed system of three equations, namely eqn (4.66) and the equivalent pairs eqns (4.67), (4.68) or eqns (4.70), (4.71), that describe the evolution of the three small perturbations $\delta\rho$, $\delta\boldsymbol{B}$, and $\boldsymbol{v}$. These equations are first-order in the differential operators $\partial/\partial t$ and $\boldsymbol{\nabla}$. We may now proceed to eliminate variables, at the price of operating again with $\boldsymbol{\nabla}$ or $\partial/\partial t$ to produce higher-order differential equations. Taking the partial time derivative of eqn (4.71), and using eqn (4.66) to eliminate $(\partial/\partial t)\delta p$, we obtain

$$\frac{\partial^2 \boldsymbol{v}}{\partial t^2} = V_\mathrm{s}^2\boldsymbol{\nabla}(\boldsymbol{\nabla}\cdot\boldsymbol{v}) - \frac{V_\mathrm{A}^2}{B_0}\left(\boldsymbol{\nabla}\frac{\partial}{\partial t}\delta B_z - \frac{\partial}{\partial z}\frac{\partial}{\partial t}\delta\boldsymbol{B}\right). \tag{4.72}$$

On the right of eqn (4.72), we can eliminate $(\partial/\partial t)\delta\boldsymbol{B}$ and $(\partial/\partial t)\delta\boldsymbol{B}_z$ using eqn (4.70). This gives an equation in terms of $\boldsymbol{v}$ alone:

$$\frac{\partial^2 \boldsymbol{v}}{\partial t^2} = (V_\mathrm{s}^2 + V_\mathrm{A}^2)\boldsymbol{\nabla}(\boldsymbol{\nabla}\cdot\boldsymbol{v}) + V_\mathrm{A}^2\left(\frac{\partial^2 \boldsymbol{v}}{\partial z^2} - \boldsymbol{\nabla}\frac{\partial v_z}{\partial z} - \hat{\boldsymbol{e}}_z\frac{\partial}{\partial z}\boldsymbol{\nabla}\cdot\boldsymbol{v}\right). \tag{4.73}$$

We establish the normal modes by taking $\boldsymbol{v}$ to vary as $\boldsymbol{v}_1 \exp(\mathrm{i}\boldsymbol{k}\cdot\boldsymbol{x} - \mathrm{i}\omega t)$, where $\boldsymbol{v}_1$ is a constant amplitude. Writing $k_\parallel$ for $\boldsymbol{k}\cdot\hat{\boldsymbol{e}}_z$, eqn (4.73) then gives

$$\{(\omega^2 - V_\mathrm{A}^2 k_\parallel^2)\boldsymbol{I} - (V_\mathrm{s}^2 + V_\mathrm{A}^2)\boldsymbol{k}\boldsymbol{k} + V_\mathrm{A}^2 k_\parallel(\boldsymbol{k}\hat{\boldsymbol{e}}_z + \hat{\boldsymbol{e}}_z\boldsymbol{k})\}\cdot\boldsymbol{v}_1 = 0. \tag{4.74}$$

Here, as usual, $\boldsymbol{I}$ denotes the identity matrix. We have not yet specified the orientation of the x-axis and y-axis in the plane perpendicular to the magnetic field. It is convenient to allow the perpendicular component of $\boldsymbol{k}$ to determine our choice of direction for the y-axis by writing

$$\boldsymbol{k} = k_\parallel\hat{\boldsymbol{e}}_z + k_\perp\hat{\boldsymbol{e}}_y. \tag{4.75}$$

Then eqn (4.74) becomes

$$\begin{bmatrix} \omega^2 - V_A^2 k_\parallel^2 & 0 & 0 \\ 0 & \omega^2 - V_A^2 k_\parallel^2 - (V_s^2 + V_A^2)k_\perp^2 & -V_s^2 k_\parallel k_\perp \\ 0 & -V_s^2 k_\parallel k_\perp & \omega^2 - V_s^2 k_\parallel^2 \end{bmatrix} \begin{bmatrix} v_{1x} \\ v_{1y} \\ v_{1z} \end{bmatrix} = 0. \tag{4.76}$$

We note first that the properties of the x-component of fluid motion are independent of the other two components. It follows that there exists a normal mode involving only fluid displacement in the x-direction, perpendicular to the yz-plane which is the common plane of $\boldsymbol{k}$ and $\boldsymbol{B}$. By eqn (4.76), its dispersion relation is

$$\omega = k_\parallel V_A. \tag{4.77}$$

This is the shear Alfvén wave. Turning to the four remaining non-zero elements of the matrix in eqn (4.76), we see that there are two normal modes that involve fluid motion in the common plane of $\boldsymbol{k}$ and $\boldsymbol{B}$. The dispersion relation follows from the vanishing of the determinant, which gives

$$\omega^4 - \omega^2(V_s^2 + V_A^2)k^2 + V_s^2 V_A^2 k^2 k_\parallel^2 = 0. \tag{4.78}$$

Using the standard formula for the roots of a quadratic equation, eqn (4.78) yields

$$\omega^2 = \frac{k^2}{2}[(V_s^2 + V_A^2) \pm \{(V_s^2 - V_A^2)^2 + 4V_s^2 V_A^2 (k_\perp^2/k^2)\}^{\frac{1}{2}}]. \tag{4.79}$$

The two distinct roots that correspond to the + and − signs in eqn (4.79) have different phase velocities. They are known as the fast and slow magnetosonic waves, respectively, and we shall discuss them further later in this section.

Now that we have identified the normal modes, it is interesting to consider the roles played by the different components of the magnetohydrodynamic force that were discussed in Section 4.2. Let us return to the force equation, eqn (4.68), and use the fact that $\boldsymbol{v}$, $\delta\boldsymbol{B}$, and δp all oscillate as $\exp(\mathrm{i}\boldsymbol{k}\cdot\boldsymbol{x} - \mathrm{i}\omega t)$. Then eqn (4.68) can be writeen

$$\omega\rho_0\boldsymbol{v} = \boldsymbol{k}\delta p + \boldsymbol{k}\left(\frac{\boldsymbol{B}_0\cdot\delta\boldsymbol{B}}{\mu_0}\right) - \delta\boldsymbol{B}\left(\frac{\boldsymbol{B}_0\cdot\boldsymbol{k}}{\mu_0}\right). \tag{4.80}$$

The perturbed, oscillating fluid velocity $\boldsymbol{v}$ has three components. First, there is the component along the magnetic field, $v_z = \hat{\boldsymbol{e}}_z\cdot\boldsymbol{v}$. Taking the scalar product of eqn (4.80) with $\hat{\boldsymbol{e}}_z$, we obtain

$$\omega\rho_0 v_z = k_\parallel \delta p, \tag{4.81}$$

where we have used the notation of eqn (4.75) for $\boldsymbol{k}$. It follows from eqn

(4.81) that oscillating fluid motion along the magnetic field is associated only with changes in fluid pressure—there are no associated perturbations of the magnetic field.

Second, there is the component of fluid motion which is parallel to the component of $\boldsymbol{k}$ that is perpendicular to the magnetic field. Again using the notation of eqn (4.75), this component is $v_y = \hat{\boldsymbol{e}}_y \cdot \boldsymbol{v}$. Then taking the scalar product of eqn (4.80) with $\hat{\boldsymbol{e}}_y$, we obtain

$$\omega \rho_0 v_y = k_\perp \delta p + \frac{B_0}{\mu_0}(k_\perp \delta B_z - k_\parallel \delta B_y). \tag{4.82}$$

Now the magnetic field must satisfy $\nabla \cdot \boldsymbol{B} = 0$, so that

$$k_\parallel \delta B_z = -k_\perp \delta B_y. \tag{4.83}$$

We therefore multiply eqn (4.82) by $k_\perp$, and then use eqn (4.83) to eliminate δB_y. This gives

$$k_\perp v_y = \frac{1}{\omega \rho_0}\left(k_\perp^2 \delta p + \frac{k^2 B_0 \delta B_z}{\mu_0}\right). \tag{4.84}$$

The quantity $B_0 \delta B_z = \boldsymbol{B}_0 \cdot \delta \boldsymbol{B} = \delta(\boldsymbol{B} \cdot \boldsymbol{B})/2 = \delta(B^2/2)$, so that eqn (4.84) can be written

$$k_\perp v_y = \frac{1}{\omega \rho_0}\left\{k_\perp^2 \delta p + k^2 \delta\left(\frac{B^2}{2\mu_0}\right)\right\}. \tag{4.85}$$

It follows that oscillation of v_y is associated both with changes in fluid pressure perpendicular to the magnetic field and with changes in magnetic field strength. Let us now combine eqns (4.81) and (4.85) to obtain the component of fluid motion that is directed parallel to the wavevector:

$$\boldsymbol{k} \cdot \boldsymbol{v} = k_\parallel v_z + k_\perp v_y = \frac{k^2}{\omega \rho_0}\delta\left(p + \frac{B^2}{2\mu_0}\right). \tag{4.86}$$

The combination $p + B^2/2\mu_0$ has previously appeared in, for example, eqn (4.45). We recall that it describes the combined pressure of the magnetic field and of the fluid. Physically, it is this combined pressure which provides the restoring force for the magnetosonic waves which, as we saw in eqns (4.76) and (4.78), involve only v_y and v_z. The fact that this restoring force has two constituents accounts in the following way for the existence of two branches, fast and slow, for the magnetosonic wave. We can set up the initial perturbations associated with a magnetosonic wave in two distinct ways. The initial values of δp and $\delta(B^2/2\mu_0)$ can have the same signs, or opposite signs. Subsequently, δp and $\delta(B^2/2\mu_0)$ will oscillate in phase, or with a phase difference of π, respectively. The maximum restoring force is greater in the former case than in the latter, which accordingly correspond respectively to the fast and slow magnetosonic waves.

Finally, we consider the third component of oscillating fluid motion, $v_x = \hat{\boldsymbol{e}}_x \cdot \boldsymbol{v}$, which is perpendicular to the plane spanned by $\boldsymbol{B}_0$ and $\boldsymbol{k}$. As we saw in eqns (4.76) and (4.77), v_x is the only velocity component that is involved in the shear Alfvén wave. Taking the scalar product of eqn (4.80) with $\hat{\boldsymbol{e}}_x$, we obtain

$$\omega \rho_0 v_x = -\delta B_x \frac{k_\parallel B_0}{\mu_0}. \tag{4.87}$$

The right-hand side of eqn (4.87) comes entirely from the term proportional to $\delta\boldsymbol{B}(\boldsymbol{B}_0 \cdot \boldsymbol{k})$ on the right-hand side of eqn (4.80). This in turn comes from the term on the right-hand side of eqn (4.68) which is proportional to $(\boldsymbol{B}_0 \cdot \nabla)\delta\boldsymbol{B}$. We examined this term in eqn (4.48), where it was divided into two parts: the first part was identified with magnetic tension; the second part is parallel to the magnetic field direction, so that its scalar product with $\hat{\boldsymbol{e}}_x$ is zero. It follows that the right-hand side of eqn (4.87), and hence the restoring force for the shear Alfvén wave, has its physical origin in the magnetic tension. To complete this discussion, we use the fact that by eqn (4.53), $V_A^2 = B_0^2/\mu_0\rho_0$. Then eqn (4.87) can be written

$$\frac{\omega}{k_\parallel} v_x = -V_A^2 \frac{\delta B_x}{B_0}. \tag{4.88}$$

We have already established in eqn (4.77) that for the shear Alfvén wave, $\omega/k_\parallel = V_A$, so that eqn (4.88) gives

$$\frac{v_x}{V_A} = -\frac{\delta B_x}{B_0}. \tag{4.89}$$

Throughout this section, we have made the linear approximation $|\delta B_x/B_0| \ll 1$. It therefore follows from eqn (4.89) that the magnitude of the fluid velocity v_x associated with a linear shear Alfvén wave is much less than V_A, the phase velocity of the wave itself.

4.4 Magnetohydrodynamic energy

At the end of Section 4.2, we identified an equilibrium configuration for a fluid governed by the equations of magnetohydrodynamics. The nature of an equilibrium is determined by the forces that arise when it is perturbed. In a stable equilibrium, these are restoring forces, which give rise to the magnetohydrodynamic oscillations that were described in Section 4.3. The equilibrium is, however, unstable if a lower energy state can be reached as a result of small perturbations. It is therefore useful to construct a general formula for magnetohydrodynamic energy.

In this general formula, we expect two classes of term to appear. First, there will be pure fluid terms, which would apply to a non-conducting fluid in the absence of a magnetic field. For example, we expect contributions from fluid kinetic energy and fluid pressure. Second, there will be terms that arise from the interaction of the electric current in the fluid with the magnetic field. These will be derived by considering the work done by the $\boldsymbol{J} \times \boldsymbol{B}$ force. In order to identify the various terms, we shall approach the magnetohydrodynamic equations in a new way. In Sections 4.2 and 4.3, we usually eliminated variables between the equations, so as to deal with a single fluid quantity; for example, eqn (4.73) describes the evolution of $\boldsymbol{v}$ alone. This approach was desirable when our aim was to identify the normal modes of the system. Now, however, we wish to display explicitly the physical origin of the various contributions to the magnetohydrodynamic energy. It would clearly be counter-productive to aim to eliminate most of the fluid variables.

Our long-term objective in this section is to obtain a formula for the way in which the magnetohydrodynamic energy of a configuration changes in response to small perturbations. To do this, it will be necessary to linearize the equations of ideal magnetohydrodynamics: that is, to write the equations in terms of small departures from equilibrium, neglecting terms that have quadratic dependence on small quantities. Before taking this step, however, it is interesting to examine how energy density and energy flux appear in the full, as opposed to the linearized, equations of ideal magnetohydrodynamics.

We know that, in general, the rate of change of energy is given by the scalar product of force with velocity. Accordingly, we start by taking the scalar product of the momentum equation, eqn (4.6), with the fluid velocity $\boldsymbol{v}$. This yields

$$\rho \boldsymbol{v} \cdot \frac{\mathrm{d}\boldsymbol{v}}{\mathrm{d}t} = \boldsymbol{v} \cdot (\boldsymbol{J} \times \boldsymbol{B}) - \boldsymbol{v} \cdot \nabla p. \tag{4.90}$$

Using Leibniz' rule for the derivative of a product, we may write

$$\rho \boldsymbol{v} \cdot \frac{\mathrm{d}\boldsymbol{v}}{\mathrm{d}t} = \frac{\mathrm{d}}{\mathrm{d}t}\left(\frac{\rho v^2}{2}\right) - \frac{v^2}{2}\frac{\mathrm{d}\rho}{\mathrm{d}t}, \tag{4.91}$$

$$\boldsymbol{v} \cdot \nabla p = \nabla \cdot (p\boldsymbol{v}) - p\nabla \cdot \boldsymbol{v}. \tag{4.92}$$

We recall from eqns (4.27) and (4.28) that, for an incompressible fluid, the second terms on the right-hand sides of eqns (4.91) and (4.92) are zero. In general, both can be expressed in terms of $\mathrm{d}p/\mathrm{d}t$ using eqns (4.29) and (4.30). Thus, eqns (4.91) and (4.92) become

$$\rho \boldsymbol{v} \cdot \frac{\mathrm{d}\boldsymbol{v}}{\mathrm{d}t} = \frac{d}{dt}\left(\frac{\rho v^2}{2}\right) - \left(\frac{\rho v^2}{2p}\right)\frac{1}{\gamma}\frac{\mathrm{d}p}{\mathrm{d}t}, \tag{4.93}$$

$$\boldsymbol{v} \cdot \nabla p = \nabla \cdot (p\boldsymbol{v}) + \frac{1}{\gamma}\frac{\mathrm{d}p}{\mathrm{d}t}. \tag{4.94}$$

Substituting eqns (4.93) and (4.94) into eqn (4.90), we obtain

$$\frac{\mathrm{d}}{\mathrm{d}t}\left(\frac{\rho v^2}{2}\right) = \boldsymbol{v}\cdot(\boldsymbol{J}\times\boldsymbol{B}) - \nabla\cdot(p\boldsymbol{v}) - \left(1 - \frac{\rho v^2}{2p}\right)\frac{1}{\gamma}\frac{\mathrm{d}p}{\mathrm{d}t}. \tag{4.95}$$

The physical origin of all three terms on the right-hand side of eqn (4.95) is clear. First, as anticipated, we have the rate of working of the $\boldsymbol{J}\times\boldsymbol{B}$ force. Secondly, we have the divergence of the flux of pressure $p\boldsymbol{v}$. This is equivalent to a flux of thermal energy, since pressure is proportional to temperature and density. The appearance of the divergence of an energy flux is to be expected. Ultimately, the conservation of energy will be expressed in the form of eqn (4.1), where the mass density ρ will be replaced by the energy density. The third term on the right in eqn (4.95) describes the effect of changing pressure, and is zero for an incompressible fluid.

Let us return to the term $\boldsymbol{v}\cdot(\boldsymbol{J}\times\boldsymbol{B})$ in eqn (4.95). By the usual properties of the triple product, we have

$$\boldsymbol{v}\cdot(\boldsymbol{J}\times\boldsymbol{B}) = -\boldsymbol{J}\cdot(\boldsymbol{v}\times\boldsymbol{B}). \tag{4.96}$$

Now recall eqn (4.11), which gives the electric field $\boldsymbol{E}$ that is induced by the motion of a perfectly conducting fluid that has velocity $\boldsymbol{v}$ with respect to $\boldsymbol{B}$. Substituting from eqn (4.11) in eqn (4.96), we obtain

$$\boldsymbol{v}\cdot(\boldsymbol{J}\times\boldsymbol{B}) = \boldsymbol{J}\cdot\boldsymbol{E}. \tag{4.97}$$

Thus, we have expressed the rate of working of the force $\boldsymbol{J}\times\boldsymbol{B}$ in an alternative, physically familiar manner. It is equal to the rate of working of the electric field $\boldsymbol{E}$, which arises from fluid motion, on the electric current $\boldsymbol{J}$ in the fluid. Using eqn (4.97), we may write eqn (4.95) in the form

$$\frac{\mathrm{d}}{\mathrm{d}t}\left(\frac{\rho v^2}{2}\right) = \boldsymbol{J}\cdot\boldsymbol{E} - \nabla\cdot(p\boldsymbol{v}) - \left(1 - \frac{\rho v^2}{2p}\right)\frac{1}{\gamma}\frac{\mathrm{d}p}{\mathrm{d}t}. \tag{4.98}$$

We now express $\boldsymbol{J}\cdot\boldsymbol{E}$ in terms of electric and magnetic fields, using the approach which is used in standard derivations of the Poynting flux. Eliminating $\boldsymbol{J}$ using eqn (4.26), we have

$$\boldsymbol{J}\cdot\boldsymbol{E} = \frac{1}{\mu_0}\boldsymbol{E}\cdot(\nabla\times\boldsymbol{B}). \tag{4.99}$$

Next, recall that the energy density of the magnetic field is $B^2/2\mu_0$. Using Faraday's law of electromagnetic induction eqn (I.4), we can express the time derivative of the field energy density as

$$\frac{\partial}{\partial t}\left(\frac{B^2}{2\mu_0}\right) = \frac{1}{\mu_0}\boldsymbol{B}\cdot\frac{\partial \boldsymbol{B}}{\partial t} = -\frac{1}{\mu_0}\boldsymbol{B}\cdot(\nabla\times\boldsymbol{E}). \tag{4.100}$$

A standard vector identity gives

$$\nabla \cdot (\boldsymbol{E} \times \boldsymbol{B}) = \boldsymbol{B} \cdot (\nabla \times \boldsymbol{E}) - \boldsymbol{E} \cdot (\nabla \times \boldsymbol{B}). \tag{4.101}$$

We substitute for the first term on the right-hand side of eqn (4.101) using eqn (4.100), and substitute for the second term using eqn (4.99). This yields

$$\nabla \cdot (\boldsymbol{E} \times \boldsymbol{B}) = -\mu_0 \frac{\partial}{\partial t}\left(\frac{B^2}{2\mu_0}\right) - \mu_0 \boldsymbol{J} \cdot \boldsymbol{E}. \tag{4.102}$$

The vector

$$\boldsymbol{S} \equiv \frac{\boldsymbol{E} \times \boldsymbol{B}}{\mu_0} \tag{4.103}$$

is known as the Poynting vector. It represents the flux of field energy, and we refer to standard textbooks on electricity and magnetism for further details. Using eqn (4.103), eqn (4.102) can be written

$$\boldsymbol{J} \cdot \boldsymbol{E} = -\nabla \cdot \boldsymbol{S} - \frac{\partial}{\partial t}\left(\frac{B^2}{2\mu_0}\right). \tag{4.104}$$

Recalling eqn (4.97), eqn (4.104) states that the energy that is used when the $\boldsymbol{J} \times \boldsymbol{B}$ force acts on the moving fluid is drawn from the magnetic field energy density and from the divergence of the Poynting vector.

Now that we have seen how energy densities and energy fluxes arise naturally from the full equations of ideal magnetohydrodynamics, let us return to our main objective. Suppose that there exists an equilibrium configuration; this requires the force defined by the right-hand side of eqn (4.6) to vanish, so that

$$0 = \boldsymbol{J}_0 \times \boldsymbol{B}_0 - \nabla p_0, \tag{4.105}$$

where subscript zero denotes an equilibrium quantity. We now perturb the equilibrium. This involves changing the fluid quantities from their equilibrium values by a small amount, so that $\boldsymbol{J} = \boldsymbol{J}_0 + \boldsymbol{J}_1$, where $|\boldsymbol{J}_1| \ll |\boldsymbol{J}_0|$, and similarly for ρ, $\boldsymbol{B}$, and p. We shall assume that the equilibrium velocity is zero, and simply write $\boldsymbol{v}$ for the velocity in the perturbed configuration. Equation (4.6) gives

$$(\rho_0 + \rho_1)\frac{\mathrm{d}\boldsymbol{v}}{\mathrm{d}t} = \boldsymbol{J}_0 \times \boldsymbol{B}_0 + \boldsymbol{J}_1 \times \boldsymbol{B}_0 + \boldsymbol{J}_0 \times \boldsymbol{B}_1 + \boldsymbol{J}_1 \times \boldsymbol{B}_1 - \nabla p_0 - \nabla p_1 \tag{4.106}$$

We now linearize eqn (4.106), neglecting the terms that have quadratic dependence on small quantities, namely $\rho_1(\mathrm{d}\boldsymbol{v}/\mathrm{d}t)$, $\rho_0 \boldsymbol{v} \cdot \nabla \boldsymbol{v}$, and $\boldsymbol{J}_1 \times \boldsymbol{B}_1$, and use eqn (4.105) to eliminate the terms that involve only equilibrium quantities. The terms that remain constitute the linearized equation of motion of ideal magnetohydrodynamics:

$$\rho_0 \frac{\partial \boldsymbol{v}}{\partial t} = \boldsymbol{J}_1 \times \boldsymbol{B}_0 + \boldsymbol{J}_0 \times \boldsymbol{B}_1 - \boldsymbol{\nabla} p_1. \tag{4.107}$$

Taking the scalar product of eqn (4.107) with $\boldsymbol{v}$, we obtain

$$\rho_0 \boldsymbol{v} \cdot \frac{\partial \boldsymbol{v}}{\partial t} = (\boldsymbol{J}_1 \times \boldsymbol{B}_0) \cdot \boldsymbol{v} + (\boldsymbol{J}_0 \times \boldsymbol{B}_1) \cdot \boldsymbol{v} - \boldsymbol{v} \cdot \boldsymbol{\nabla} p_1. \tag{4.108}$$

While this expression is less general than the full expression, eqn (4.90), it is better adapted to calculating the relative energy of neighbouring magnetohydrodynamic configurations.

We begin our analysis of eqn (4.108) by noting that the fluid motion relative to the magnetic field creates an electric field

$$\boldsymbol{E} = -\boldsymbol{v} \times \boldsymbol{B}_0; \tag{4.109}$$

this expression is obtained by linearizing eqn (4.11), so that the contribution $-\boldsymbol{v} \times \boldsymbol{B}_1$ is neglected. Using the cyclic properties of the triple product, as well as eqn (4.109), the first term on the right-hand side of eqn (4.108) can be written

$$\boldsymbol{v} \cdot (\boldsymbol{J}_1 \times \boldsymbol{B}_0) = \boldsymbol{J}_1 \cdot \boldsymbol{E}. \tag{4.110}$$

This is the linearized analogue of eqn (4.97); the steps taken between eqn (4.99) and eqn (4.104) can be repeated for the present linearized variables, giving

$$\boldsymbol{J}_1 \cdot \boldsymbol{E} = -\boldsymbol{\nabla} \cdot \boldsymbol{S} - \frac{\partial}{\partial t}\left(\frac{B_1^2}{2\mu_0}\right), \tag{4.111}$$

where now the Poynting vector

$$\boldsymbol{S} = \frac{\boldsymbol{E} \times \boldsymbol{B}_1}{\mu_0}. \tag{4.112}$$

Returning to eqn (4.108), we transform the final term using Leibniz' rule, and use eqns (4.110) and (4.111) to write

$$\frac{\partial}{\partial t}(\tfrac{1}{2}\rho_0 v^2) = -\boldsymbol{\nabla} \cdot \boldsymbol{S} - \frac{\partial}{\partial t}\left(\frac{B_1^2}{2\mu_0}\right) - \boldsymbol{\nabla} \cdot (p_1 \boldsymbol{v}) + p_1 \boldsymbol{\nabla} \cdot \boldsymbol{v} + \boldsymbol{v} \cdot (\boldsymbol{J}_0 \times \boldsymbol{B}_1). \tag{4.113}$$

Our objective is to write all the terms as partial time derivatives or as divergences of fluxes. We can re-write the penultimate term in eqn (4.113) using the linearized version of eqn (4.30), which is

$$\frac{\partial p_1}{\partial t} + \boldsymbol{v} \cdot \boldsymbol{\nabla} p_0 = -\gamma p_0 \boldsymbol{\nabla} \cdot \boldsymbol{v}. \tag{4.114}$$

It follows from this equation that

$$p_1 \nabla \cdot \boldsymbol{v} = -\frac{\partial}{\partial t}\left(\frac{p_1^2}{2\gamma p_0}\right) - \frac{p_1}{2\gamma p_0}(\boldsymbol{v} \cdot \nabla)p_0. \tag{4.115}$$

Substituting eqn (4.115) into eqn (4.113), and re-arranging, we have

$$\frac{\partial}{\partial t}\left\{\tfrac{1}{2}\rho_0 v^2 + \frac{B_1^2}{2\mu_0} + \frac{p_1^2}{2\gamma p_0}\right\} + \nabla \cdot (\boldsymbol{S} + p_1 \boldsymbol{v})$$

$$= \boldsymbol{v} \cdot (\boldsymbol{J}_0 \times \boldsymbol{B}_1) - \frac{p_1}{\gamma p_0}(\boldsymbol{v} \cdot \nabla)p_0. \tag{4.116}$$

In order to progress further, we need to express the right-hand side of eqn (4.116) in terms of some new variable which can be written as a time derivative. An obvious candidate is the displacement vector field, which describes the distance and direction that the plasma has moved from equilibrium; it is, after all, our objective to express magnetohydrodynamic energy in terms of the displacement. Writing this quantity as $\boldsymbol{\xi}$, we have, by definition,

$$\boldsymbol{v} = \frac{\mathrm{d}\boldsymbol{\xi}}{\mathrm{d}t} = \frac{\partial \boldsymbol{\xi}}{\partial t} + (\boldsymbol{v} \cdot \nabla)\boldsymbol{\xi}. \tag{4.117}$$

Linearizing, as usual, we neglect the final term in eqn (4.117). Then, for example, the linearized version of eqn (4.13),

$$\frac{\partial \boldsymbol{B}_1}{\partial t} = \nabla \times (\boldsymbol{v} \times \boldsymbol{B}_0) \tag{4.118}$$

becomes

$$\frac{\partial \boldsymbol{B}_1}{\partial t} = \nabla \times \left(\frac{\partial \boldsymbol{\xi}}{\partial t} \times \boldsymbol{B}_0\right), \tag{4.119}$$

and hence

$$\boldsymbol{B}_1 = \nabla \times (\boldsymbol{\xi} \times \boldsymbol{B}_0). \tag{4.120}$$

Returning to eqn (4.116), we may use these new expressions to write

$$\boldsymbol{v} \cdot (\boldsymbol{J}_0 \times \boldsymbol{B}_1) = -\boldsymbol{J}_0 \cdot (\boldsymbol{v} \times \boldsymbol{B}_1) = -\boldsymbol{J}_0 \cdot \left(\frac{\partial \boldsymbol{\xi}}{\partial t} \times \boldsymbol{B}_1\right), \tag{4.121}$$

$$\frac{p_1}{\gamma p_0}(\boldsymbol{v} \cdot \nabla)p_0 = \frac{p_1}{\gamma p_0}\left(\frac{\partial \boldsymbol{\xi}}{\partial t} \cdot \nabla\right)p_0. \tag{4.122}$$

In eqn (4.121), $\boldsymbol{B}_1$ is itself a linear function of $\boldsymbol{\xi}$ given by eqn (4.120); similarly, in eqn (4.122), p_1 is itself a linear function of $\boldsymbol{\xi}$, since eqn (4.114) yields

$$p_1 = -\boldsymbol{\xi} \cdot \nabla p_0 - \gamma p_0 \nabla \cdot \boldsymbol{\xi}. \tag{4.123}$$

We must not allow the apparent complexity of eqns (4.121) and (4.122)—involving, as they do, vector products and the ∇ operator—to obscure a simple fact which follows from the previous remark. For the purposes of illustration, let us for a moment replace $\boldsymbol{\xi}$ by the scalar ξ. It then follows from eqn (4.123) that $p_1 = \hat{a}\xi$, where $\hat{a}$ is a linear operator; thus the combination $p_1(\partial\xi/\partial t)$, whose vector counterpart appears in eqn (4.122), can be written

$$p_1 \frac{\partial\xi}{\partial t} = \hat{a}\xi \frac{\partial\xi}{\partial t} = \frac{1}{2}\hat{a}\xi \frac{\partial\xi}{\partial t} + \frac{1}{2}\hat{a}\frac{\partial\xi}{\partial t}\xi$$

$$= \frac{1}{2}p_1 \frac{\partial\xi}{\partial t} + \frac{1}{2}\frac{\partial p_1}{\partial t}\xi$$

$$= \frac{1}{2}\frac{\partial}{\partial t}(p_1\xi). \qquad (4.124)$$

The same line of argument may be applied to eqn (4.121), so that

$$\frac{\partial\boldsymbol{\xi}}{\partial t} \times \boldsymbol{B}_1 = \frac{1}{2}\frac{\partial}{\partial t}(\boldsymbol{\xi} \times \boldsymbol{B}_1). \qquad (4.125)$$

Substituting eqns (4.125) and (4.124) into eqns (4.121) and (4.122) respectively, and using the resulting expressions in eqn (4.116), we obtain

$$\frac{\partial}{\partial t}\left\{\frac{1}{2}\rho_0 v^2 + \frac{B_1^2}{2\mu_0} + \frac{p_1^2}{2\gamma p_0} + \frac{\boldsymbol{J}_0}{2}\cdot(\boldsymbol{\xi}\times\boldsymbol{B}_1) + \frac{p_1}{2\gamma p_0}(\boldsymbol{\xi}\cdot\nabla)p_0\right\}$$
$$+ \nabla\cdot(\boldsymbol{S} + p_1\boldsymbol{v}) = 0. \qquad (4.126)$$

This is the continuity equation for energy in linearized ideal magnetohydrodynamics. The term involving the partial time derivative describes fluid kinetic energy, magnetic field energy, plasma thermal energy, the rate of working arising from perturbed motion in the presence of perturbed magnetic fields and of an equilibrium current $\boldsymbol{J}_0$, and the effect of displacement with respect to the equilibrium pressure gradient; the remaining terms are the Poynting flux and the flux of plasma thermal energy.

Note that in eqn (4.126), $\boldsymbol{B}_1$ and p_1 can be written in terms of $\boldsymbol{\xi}$ and equilibrium quantities using eqns (4.120) and (4.123), if necessary. For example,

$$\frac{p_1^2}{2\gamma p_0} + \frac{p_1}{2\gamma p_0}(\boldsymbol{\xi}\cdot\nabla)p_0 = -\frac{p_1}{2}\nabla\cdot\boldsymbol{\xi}$$

$$= \frac{1}{2}(\nabla\cdot\boldsymbol{\xi})(\boldsymbol{\xi}\cdot\nabla)p_0 + \frac{\gamma p_0}{2}(\nabla\cdot\boldsymbol{\xi})^2. \qquad (4.127)$$

Substituting this expression into eqn (4.126) and integrating over the plasma volume, we obtain

$$\frac{\partial}{\partial t}(\delta K + \delta W) = 0, \tag{4.128}$$

where

$$\delta K = \int \tfrac{1}{2}\rho_0 v^2 \, \mathrm{d}^3\boldsymbol{x} \tag{4.129}$$

and

$$\delta W = \int \frac{1}{2}\left\{\frac{B_1^2}{\mu_0} + \boldsymbol{J}_0 \cdot (\boldsymbol{\xi} \times \boldsymbol{B}_1) + (\boldsymbol{\nabla} \cdot \boldsymbol{\xi})[(\boldsymbol{\xi} \cdot \boldsymbol{\nabla})p_0 + \gamma p_0 \boldsymbol{\nabla} \cdot \boldsymbol{\xi}]\right\} \mathrm{d}^3\boldsymbol{x}. \tag{4.130}$$

The flux terms in eqn (4.126) make no contribution to eqn (4.128) since we choose the plasma boundary to be such that, on applying the divergence theorem eqn (I.2), there is no flux across the boundary. Equation (4.128) represents our final objective in this chapter: the energy principle for linearized ideal magnetohydrodynamics. We identify δK as the kinetic energy and δW, which is a function only of $\boldsymbol{\xi}$ and equilibrium quantities, as the change in magnetohydrodynamic potential energy. If $\boldsymbol{B}_0$, $\boldsymbol{J}_0$, and p_0 are such that a displacement $\boldsymbol{\xi}$ can be found which makes δW negative, it follows from eqn (4.128) that the plasma can move from its equilibrium configuration, acquiring kinetic energy as it does so. Conversely, if no such $\boldsymbol{\xi}$ can be found, the plasma equilibrium is stable.

Exercises

4.1. Throughout this chapter, we have dealt with ideal magnetohydrodynamics, which applies to perfectly conducting fluids. The subject known as non-ideal magnetohydrodynamics applies to fluids whose conductivity σ is not infinite or, equivalently, whose resistivity is not zero.

(a) Find the equation of non-ideal magnetohydrodynamics that replaces eqn (4.13).

(b) Show that, when the evolution of the magnetic field is dominated by the new, non-ideal term, the field decays in a way described by a diffusion equation. What is the relation between the characteristic timescale and lengthscale of this process?

(c) Show that the decay of the magnetic field is due to the dissipation of energy through Joule heating.

4.2. The solar wind consists of a diffuse plasma that streams outwards from the sun and fills interplanetary space. Its density and velocity near the Earth fluctuate in time; for our purposes, we may take $n_{\mathrm{SW}} = 4 \times 10^6$ protons per cubic metre, and $v_{\mathrm{SW}} = 4 \times 10^5 \,\mathrm{m\,s^{-1}}$. The Earth's magnetic field is sufficiently strong to deflect the solar wind, and the resulting boundary between the interplanetary plasma and the terrestrial plasma (or magnetosphere) is known as the magnetopause. Taking the strength of the Earth's magnetic field in the equatorial plane to be $B = 3 \times 10^{-5}/R^3$ Tesla, where R is the distance from the centre of the Earth, measured in Earth radii, estimate the distance to the magnetopause in the direction towards the sun.

4.3. Consider a cylindrical plasma, extending along the z-axis, which carries a steady electric current whose density can be written $J(r)\hat{\mathbf{e}}_z$, where r denotes distance from the z-axis.

(a) Obtain the magnetic field as a function of r.

(b) Obtain an expression for $J(r)$ in terms of the magnetic field and its gradient.

(c) Show that the $\mathbf{J} \times \mathbf{B}$ force is directed radially inward, and has magnitude

$$\frac{1}{2\mu_0 r^2}\frac{\mathrm{d}}{\mathrm{d}r}\{r^2B^2(r)\}.$$

(d) If the plasma is in magnetohydrodynamic equilibrium, obtain an expression for the pressure $p(r)$, assuming that the pressure vanishes at the plasma boundary $r = a$.

Solutions are on pages 150 *to* 153.

5

Kinetic description of plasma

5.1 Introduction to kinetic theory

Let us first look back at the different approaches that we have used for describing plasmas. After emphasizing the collective nature of plasma behaviour in Chapter 1, we considered the dynamics of single particles in Chapter 2. In order to achieve a collective description, we needed to sum the effects of single-particle dynamics for all the particles in the plasma. This summation was carried out in a simplistic, but nevertheless successful, way at eqn (3.21). By considering the current created by the sum of the individual electron velocities, we constructed the dielectric tensor of the plasma. This tensor was then employed in Maxwell's equations to yield the normal modes of the plasma. By including the contribution of the ions, this approach was extended to low frequencies. The low-frequency description of plasmas was developed further in Chapter 4, using a macroscopic approach based on fluid quantities, such as the bulk velocity $\boldsymbol{v}$. A closed set of coupled equations that describe the evolution of these bulk quantities was introduced a priori, rather than deduced from single-particle considerations. This magnetohydrodynamic description is successful in two respects: the normal modes that it yields are identical, where appropriate, to those obtained from the dielectric approach of Chapter 3; also, magnetohydrodynamics gives a description, at a macroscopic level, of the forces that act on a plasma and the way in which a plasma moves. This goes beyond what is available from the dielectric description in Chapter 3.

We can now select two important concepts from Chapters 3 and 4: the summation of the effects of the dynamics of individual microscopic particles, as at eqn (3.21); and the continuity of the plasma when considered as a macroscopic fluid, expressed by eqn (4.1). Both concepts have already proved useful in isolation from each other. In kinetic theory, they are combined to give a powerful description of the plasma.

First, we note that consideration of the dynamics of a single particle must, by definition, involve its velocity $\boldsymbol{v}$ as well as its position $\boldsymbol{x}$. In order to specify the state of a single particle so that we can calculate the forces that are acting on it, we need to know the values of t, $\boldsymbol{x}$, and $\boldsymbol{v}$. By contrast, in order to calculate the rates of change of the macroscopic quantities considered in Chapter 4, we need only t and $\boldsymbol{x}$. Thus, in kinetic theory, the explicit inclusion of single particle dynamics means that we expect to see $\boldsymbol{x}$, $\boldsymbol{v}$, and t as the basic independent variables. The six coordinates contained in $\boldsymbol{x}$ and $\boldsymbol{v}$ are said to lie in phase space.

Next, we need to establish the identity of the fundamental quantity in terms of which kinetic theory will be developed. So far, we know only that it should have $(\boldsymbol{x}, \boldsymbol{v}, t)$ as its arguments. The physical fact that determines the identity of this quantity is as follows: the total number N of particles does not change, regardless of the way in which individual particles move in phase space. This is the basic continuity property of the system. It is analogous to the basic continuity property of the fluid considered in Chapter 4, namely the conservation of total fluid mass described by eqn (4.1). Now eqn (4.1) involves the mass density in space, $\rho(\boldsymbol{x}, t)$. The kinetic analogue of eqn (4.1), which must conserve number rather than mass, should therefore involve the number density in phase space, which we write as $f(\boldsymbol{x}, \boldsymbol{v}, t)$. The meaning of f is analogous to that of ρ, which we now recall. Consider a volume element $\mathrm{d}^3\boldsymbol{x}$, which at time t is located at position $\boldsymbol{x}$. The definition of the mass density ρ is that the mass of fluid contained within the volume element at position $\boldsymbol{x}$ at time t is

$$\mathrm{d}M(\boldsymbol{x}, t) = \rho(\boldsymbol{x}, t)\,\mathrm{d}^3\boldsymbol{x}. \tag{5.1}$$

Our analogue to $\mathrm{d}M(\boldsymbol{x}, t)$ is $\mathrm{d}N(\boldsymbol{x}, \boldsymbol{v}, t)$: the number of particles whose position lies within the small volume element $\mathrm{d}^3\boldsymbol{x}$ at position $\boldsymbol{x}$, and whose velocity lies within the velocity space element $\mathrm{d}^3\boldsymbol{v}$ at $\boldsymbol{v}$, at time t. Then $f(\boldsymbol{x}, \boldsymbol{v}, t)$ is defined by

$$\mathrm{d}N(\boldsymbol{x}, \boldsymbol{v}, t) = f(\boldsymbol{x}, \boldsymbol{v}, t)\,\mathrm{d}^3\boldsymbol{x}\,\mathrm{d}^3\boldsymbol{v}. \tag{5.2}$$

As required, $f(\boldsymbol{x}, \boldsymbol{v}, t)$ is the number density of particles in phase space. It is referred to as the distribution function of the particles. If we integrate $f(\boldsymbol{x}, \boldsymbol{v}, t)$ over the entire spatial volume that is occupied by the plasma, we obtain a function which describes the density in velocity space of the total population of plasma particles. This is the familiar velocity distribution of the particles at time t. If we go on to integrate over all velocities as well, we obtain

$$\iint f(\boldsymbol{x}, \boldsymbol{v}, t)\,\mathrm{d}^3\boldsymbol{x}\,\mathrm{d}^3\boldsymbol{v} = N, \tag{5.3}$$

the total number of particles in the plasma. Thus eqn (5.3) defines the normalization of $f(\boldsymbol{x}, \boldsymbol{v}, t)$, and is analogous to eqn (4.2).

Thus far, we have seen that kinetic theory, which must reflect both single-particle dynamics and the overall continuity of the plasma, should be developed in terms of the plasma distribution function $f(\boldsymbol{x}, \boldsymbol{v}, t)$. We expect this number density in phase space to obey an equation that is analogous to eqn (4.1). Now eqn (4.1) has two terms: the partial time derivative of the density ρ, and the three-dimensional divergence of the three-dimensional flux $\rho \boldsymbol{v}$. Let us write eqn (4.1) in the form

$$\frac{\partial \rho}{\partial t} + \frac{\partial}{\partial \boldsymbol{x}} \cdot \left(\rho \frac{\mathrm{d}\boldsymbol{x}}{\mathrm{d}t}\right) = 0. \tag{5.4}$$

We have established that f is analogous to ρ. Since phase space is six-dimensional, the flux of particles in phase space has six components: $(f\,\mathrm{d}\boldsymbol{x}/\mathrm{d}t,\, f\,\mathrm{d}\boldsymbol{v}/\mathrm{d}t)$. It is therefore appropriate to consider the six-dimensional divergence of this six-dimensional flux. Following the model of eqn (5.4), we write our continuity equation as

$$\frac{\partial f}{\partial t} + \frac{\partial}{\partial \boldsymbol{x}} \cdot \left(f \frac{\mathrm{d}\boldsymbol{x}}{\mathrm{d}t}\right) + \frac{\partial}{\partial \boldsymbol{v}} \cdot \left(f \frac{\mathrm{d}\boldsymbol{v}}{\mathrm{d}t}\right) = 0. \tag{5.5}$$

The discussion leading to eqn (4.3) applies again here, generalized to six-dimensional phase space, and need not be repeated. We note that

$$\frac{\partial}{\partial \boldsymbol{x}} \cdot \left(f \frac{\mathrm{d}\boldsymbol{x}}{\mathrm{d}t}\right) = \frac{\partial}{\partial \boldsymbol{x}} \cdot (f\boldsymbol{v}) = \boldsymbol{v} \cdot \frac{\partial f}{\partial \boldsymbol{x}} + f \frac{\partial}{\partial \boldsymbol{x}} \cdot \boldsymbol{v} = \boldsymbol{v} \cdot \frac{\partial f}{\partial \boldsymbol{x}}. \tag{5.6}$$

The quantity $\dfrac{\partial}{\partial \boldsymbol{x}} \cdot \boldsymbol{v}$ vanishes, since $\boldsymbol{v}$ and $\boldsymbol{x}$ are independent coordinates of equal standing—$\boldsymbol{v}$ is not a function of $\boldsymbol{x}$. Thus eqn (5.5) simplifies to

$$\frac{\partial f}{\partial t} + \boldsymbol{v} \cdot \frac{\partial f}{\partial \boldsymbol{x}} + \frac{\mathrm{d}\boldsymbol{v}}{\mathrm{d}t} \cdot \frac{\partial f}{\partial \boldsymbol{v}} + f \frac{\partial}{\partial \boldsymbol{v}} \cdot \frac{\mathrm{d}\boldsymbol{v}}{\mathrm{d}t} = 0. \tag{5.7}$$

Equation (5.7) is a general equation that describes the continuity of a system of particles in phase space. So far, it contains nothing that relates it specifically to plasmas. Also, we have not yet included any specific type of particle dynamics. Clearly, the next step is to introduce plasma particle dynamics explicitly. As usual, we employ the Lorentz force eqn (I.16),

$$\frac{\mathrm{d}\boldsymbol{v}}{\mathrm{d}t} = \frac{q}{m}(\boldsymbol{E}(\boldsymbol{x}, t) + \boldsymbol{v} \times \boldsymbol{B}(\boldsymbol{x}, t)). \tag{5.8}$$

We note that $\mathrm{d}\boldsymbol{v}/\mathrm{d}t$ given by eqn (5.8) is a function of all seven coordinates $\boldsymbol{x}$, $\boldsymbol{v}$, and t. Also $\mathrm{d}v_x/\mathrm{d}t$, $\mathrm{d}v_y/\mathrm{d}t$, and $\mathrm{d}v_z/\mathrm{d}t$ are independent of v_x, v_y, and v_z respectively, so that the quantity $(\partial/\partial \boldsymbol{v}) \cdot (\mathrm{d}\boldsymbol{v}/\mathrm{d}t)$ that arises in eqn (5.7) is zero. It follows that for a plasma, the equation of continuity in phase space is

$$\frac{\partial f}{\partial t}(\boldsymbol{x}, \boldsymbol{v}, t) + \boldsymbol{v} \cdot \frac{\partial f}{\partial \boldsymbol{x}}(\boldsymbol{x}, \boldsymbol{v}, t) + \frac{q}{m}\{\boldsymbol{E}(\boldsymbol{x}, t) + \boldsymbol{v} \times \boldsymbol{B}(\boldsymbol{x}, t)\} \cdot \frac{\partial f}{\partial \boldsymbol{v}}(\boldsymbol{x}, \boldsymbol{v}, t) = 0. \tag{5.9}$$

This is referred to as the Vlasov equation. We note that it applies separately to each species of particle (electrons and ions) in the plasma.

It is now interesting to derive a simple general theorem concerning the Vlasov equation and its solutions. Suppose that the macroscopic fields $\boldsymbol{E}$ and $\boldsymbol{B}$ are given (we shall discuss self-consistency shortly). Then, following Newton's laws of motion, the position and velocity of each particle are completely determined at all times by its initial position and initial velocity. Denoting these six constants for a given particle by $(\gamma_1, \ldots, \gamma_6)$, we may therefore write the position and velocity of each particle as functions of these constants and of time:

$$\boldsymbol{x} = \boldsymbol{x}(\gamma_1, \ldots, \gamma_6, t), \qquad \boldsymbol{v} = \boldsymbol{v}(\gamma_1, \ldots, \gamma_6, t). \tag{5.10}$$

The actual form of the functional dependence on $(\gamma_1, \ldots, \gamma_6, t)$ is determined by the structure of $\boldsymbol{E}$ and $\boldsymbol{B}$. Equation (5.10) relates the six components of $(\boldsymbol{x}, \boldsymbol{v})$ to the six constants $(\gamma_1, \ldots, \gamma_6)$ at given t. This relation can in principle be inverted, so that formally we may write the constants as functions of $(\boldsymbol{x}, \boldsymbol{v}, t)$:

$$\gamma_1 = \gamma_1(\boldsymbol{x}, \boldsymbol{v}, t), \ldots, \gamma_6 = \gamma_6(\boldsymbol{x}, \boldsymbol{v}, t). \tag{5.11}$$

The values of $\gamma_1, \ldots, \gamma_6$ do not change as $\boldsymbol{x}$ and $\boldsymbol{v}$ change, following the particle motion. Now consider a distribution function where $\boldsymbol{x}$, $\boldsymbol{v}$, and t enter only through the functional combinations given in eqn (5.11): that is, any distribution function of the form

$$f(\boldsymbol{x}, \boldsymbol{v}, t) = g(\gamma_1, \ldots, \gamma_6). \tag{5.12}$$

Substituting eqn (5.12) into the Vlasov equation eqn (5.9), we obtain

$$\sum_{i=1}^{6} \frac{\partial g}{\partial \gamma_i} \left\{ \frac{\partial \gamma_i}{\partial t} + \boldsymbol{v} \cdot \frac{\partial \gamma_i}{\partial \boldsymbol{x}} + \frac{q}{m} (\boldsymbol{E} + \boldsymbol{v} \times \boldsymbol{B}) \cdot \frac{\partial \gamma_i}{\partial \boldsymbol{v}} \right\} = 0. \tag{5.13}$$

Now the quantity in { } in eqn (5.13) is simply the rate of change of γ_i following the particle motion in phase space, since $\partial \gamma_i / \partial \boldsymbol{x}$ is combined with $\mathrm{d}\boldsymbol{x}/\mathrm{d}t$ and $\partial \gamma_i / \partial \boldsymbol{v}$ is combined with $\mathrm{d}\boldsymbol{v}/\mathrm{d}t$. By definition, γ_i is a constant of the particle motion, so that this rate of change is always zero. Thus eqn (5.13) is satisfied identically, regardless of the nature of the $\partial g / \partial \gamma_i$ terms. This is Jeans' theorem: any function of the constants of motion of a particle is a solution of the Vlasov equation.

Jeans' theorem has immediate practical consequences. For example, if there are no electric and magnetic fields, $mv^2/2$ is a constant of the motion. Then any function of $mv^2/2$ is a solution of the Vlasov equation, in particular the familiar $f(v) = \exp(-mv^2/2k_\mathrm{B}T)$. Next, suppose that there is an electrostatic field $\boldsymbol{E} = -\boldsymbol{\nabla}\phi(\boldsymbol{x})$. Then $mv^2/2 + q\phi(\boldsymbol{x})$ is a constant of the motion, and $f(\boldsymbol{x}, \boldsymbol{v}) = \exp(-mv^2/2k_\mathrm{B}T - q\phi/k_\mathrm{B}T)$ is a solution of the Vlasov equation. We have already seen this type of distribution in Section 1.2, where we used it to derive the Debye length. Previously, the distribution was justified using dynamical arguments at eqns (1.16) and (1.17). We have now seen how it arises naturally in kinetic theory using Jeans' theorem.

The final step in setting up plasma kinetic theory is to construct a closed self-consistent set of equations. In eqn (5.9), the quantities $\boldsymbol{E}(\boldsymbol{x}, t)$ and $\boldsymbol{B}(\boldsymbol{x}, t)$ are macroscopic fields that act on all particles at $(\boldsymbol{x}, t)$. They do not include microscopic fields associated with binary collisions, which will be included within kinetic theory in a later section. Thus $\boldsymbol{E}$ and $\boldsymbol{B}$ in eqn (5.9) have two types of origin. First, they may be applied from outside the system: these we shall ignore. Second, fields may be generated by the charge density and current density within the plasma: these can be calculated by integrating the local values of f and of $\boldsymbol{v}f$ over all the velocities, and summing over all the species of plasma particle. Denoting the different species by subscript α, Maxwell's equations eqns. (I.1) and (I.9) become

$$\nabla \cdot \boldsymbol{E}(\boldsymbol{x}, t) = \sum_{\alpha} \frac{q_\alpha}{\varepsilon_0} \int f_\alpha(\boldsymbol{x}, \boldsymbol{v}, t)\, \mathrm{d}^3\boldsymbol{v}, \tag{5.14}$$

$$\nabla \times \boldsymbol{H}(\boldsymbol{x}, t) = \sum_{\alpha} q_\alpha \int \boldsymbol{v} f_\alpha(\boldsymbol{x}, \boldsymbol{v}, t)\, \mathrm{d}^3\boldsymbol{v} + \varepsilon_0 \frac{\partial \boldsymbol{E}}{\partial t}(\boldsymbol{x}, t). \tag{5.15}$$

Here, the familiar operator ∇ is, of course, identical to the operator which we have written as $\partial/\partial \boldsymbol{x}$ in eqn (5.9).

This completes our derivation of the basic equations of plasma kinetic theory. These equations comprise the Vlasov equation, eqn (5.9), together with the source equations for $\boldsymbol{E}$ and $\boldsymbol{H}$ expressed in terms of the plasma distribution function, eqns (5.14) and (5.15). We shall now examine how this closed system of integro-differential equations yields further information on plasma dynamics.

5.2 Kinetic theory of unmagnetized plasma

We shall consider a plasma that is initially spatially homogeneous and in equilibrium, and that is not subject to an externally applied magnetic field. For simplicity, we shall assume that the ions in the plasma remain immobile on the timescale of interest, because of their inertia, so that only perturbations of the electron distribution function are significant. The equilibrium electron distribution function f_0 is static, so that $\partial f_0/\partial t$ is zero. If the equilibrium plasma is spatially homogeneous, f_0 must be also be independent of $\boldsymbol{x}$. Thus f_0 is a function only of $\boldsymbol{v}$. At equilibrium, there are no electric fields in the plasma. If the electron and ion charges are equal and opposite, it follows from eqn (5.14) that the equilibrium electron and ion number densities in real space are equal. Denoting this number density by n_0, we follow eqn (5.3) and define the normalization of f_0 by

$$\int f_0(\boldsymbol{v})\, \mathrm{d}^3\boldsymbol{v} = n_0. \tag{5.16}$$

Now let us consider small perturbations $f_1(\boldsymbol{x}, \boldsymbol{v}, t)$ of the electron distribution function, with $f_1 \ll f_0$. The existence of this perturbation will give rise to an electric field within the plasma. By eqn (5.14),

$$\nabla \cdot \boldsymbol{E} = -\frac{e}{\varepsilon_0} \int f_1 \, \mathrm{d}^3 \boldsymbol{v}. \tag{5.17}$$

Here, we have written the electron charge explicitly as $-e$. This electric field will in turn accelerate the plasma electrons, and acts as a driving term in the Vlasov equation, eqn (5.9). If we neglect terms that are quadratic in the small quantity f_1, we obtain the linearized Vlasov equation for an unmagnetized plasma:

$$\frac{\partial f_1}{\partial t} + \boldsymbol{v} \cdot \nabla f_1 - \frac{e}{m} \boldsymbol{E} \cdot \frac{\partial f_0}{\partial \boldsymbol{v}} = 0. \tag{5.18}$$

Given a functional form for f_0, eqns (5.17) and (5.18) represent two equations in two unknowns, f_1 and $\boldsymbol{E}$. The Vlasov equation has given an independent relation between f_1 and $\boldsymbol{E}$, additional to that which follows from Poisson's equation. This has enabled us to construct a closed, self-consistent system of equations. To facilitate its solution, it is convenient to express the dependence on $\boldsymbol{x}$ and t of f_1 and $\boldsymbol{E}$ as the sum of harmonic terms, each with its own amplitude. The amplitudes of $f_1(\boldsymbol{x}, \boldsymbol{v}, t)$ and $\boldsymbol{E}(\boldsymbol{x}, t)$ that oscillate as $\exp\{\mathrm{i}(\boldsymbol{k} \cdot \boldsymbol{x} \cdot \omega t)\}$ will be written as $f_{1\boldsymbol{k}\omega}(\boldsymbol{v})$ and $\boldsymbol{E}_{\boldsymbol{k}\omega}$, respectively; note that we have not made any constraint on the values of $\boldsymbol{k}$ and ω. By eqns (5.17) and (5.18), the amplitudes of particular harmonic components of f_1 and $\boldsymbol{E}$ are related in two ways:

$$\mathrm{i}\boldsymbol{k} \cdot \boldsymbol{E}_{\boldsymbol{k}\omega} = -\frac{e}{\varepsilon_0} \int f_{1\boldsymbol{k}\omega}(\boldsymbol{v}) \, \mathrm{d}^3 \boldsymbol{v}, \tag{5.19}$$

$$-\mathrm{i}\omega f_{1\boldsymbol{k}\omega}(\boldsymbol{v}) + \mathrm{i}\boldsymbol{k} \cdot \boldsymbol{v} f_{1\boldsymbol{k}\omega}(\boldsymbol{v}) - \frac{e}{m} \boldsymbol{E}_{\boldsymbol{k}\omega} \cdot \frac{\partial f_0}{\partial \boldsymbol{v}} = 0. \tag{5.20}$$

It follows from eqn (5.20) that

$$f_{1\boldsymbol{k}\omega}(\boldsymbol{v}) = \frac{\mathrm{i}e}{m} \frac{1}{\omega - \mathbf{k} \cdot \boldsymbol{v}} \boldsymbol{E}_{\boldsymbol{k}\omega} \cdot \frac{\partial f_0}{\partial \boldsymbol{v}}. \tag{5.21}$$

Substituting this expression into the integral in eqn (5.19), we obtain

$$\boldsymbol{k} \cdot \boldsymbol{E}_{\boldsymbol{k}\omega} = -\frac{e^2}{\varepsilon_0 m} \boldsymbol{E}_{\boldsymbol{k}\omega} \cdot \int \frac{\partial f_0 / \partial \boldsymbol{v}}{\omega - \boldsymbol{k} \cdot \boldsymbol{v}} \mathrm{d}^3 \boldsymbol{v}. \tag{5.22}$$

If there is no magnetic field associated with the wave, the remaining Maxwell's equations show that, as usual for an electrostatic wave, $\boldsymbol{k}$, $\boldsymbol{E}$, and $\boldsymbol{J}$ all have the same orientation in space. Let us choose our spatial z-axis to lie along this direction, so that the component of $\boldsymbol{v}$ along it is denoted by v_z. Then if $\boldsymbol{E}_{\boldsymbol{k}\omega}$ is to be non-zero, eqn (5.22) requires

$$1+\frac{e^2}{\varepsilon_0 mk}\int\frac{\partial f_0/\partial v_z}{\omega-kv_z}\,\mathrm{d}v_z\,\mathrm{d}v_x\,\mathrm{d}v_y=0. \tag{5.23}$$

Equation (5.23) implicitly defines a relation between ω and k—that is, a dispersion relation—which depends on the functional form of $f_0(\boldsymbol{v})$. Let us first consider a cold plasma. In this case, it is convenient to use the fact that

$$\frac{\partial}{\partial v_z}\left(\frac{f_0}{\omega-\boldsymbol{k}\cdot\boldsymbol{v}}\right)=\frac{\partial f_0/\partial v_z}{\omega-\boldsymbol{k}\cdot\boldsymbol{v}}+\frac{kf_0}{(\omega-\boldsymbol{k}\cdot\boldsymbol{v})^2}. \tag{5.24}$$

If we integrate eqn (5.24) with respect to v_z from $v_z=-\infty$ to $v_z=\infty$, the left-hand side vanishes, both because f_0 tends to zero at $v_z=\pm\infty$, and because kv_z tends to $\pm\infty$ at $v_z=\pm\infty$. We can therefore write eqn (5.23) in the form

$$1-\frac{e^2}{\varepsilon_0 m}\int\frac{f_0}{(\omega-kv_z)^2}\,\mathrm{d}v_z\,\mathrm{d}v_x\,\mathrm{d}v_y=0. \tag{5.25}$$

Now in a cold plasma, by definition, f_0 is zero everywhere except when $\boldsymbol{v}=0$. Therefore the denominator $(\omega-kv_z)^2$ only matters when $v_z=0$, when it takes the value ω^2. Using the normalization of f_0 given by eqn (5.16), eqn (5.25) gives

$$1-\frac{n_0e^2}{\varepsilon_0 m\omega^2}=0=1-\frac{\omega_{\mathrm{pe}}^2}{\omega^2}, \tag{5.26}$$

for a cold plasma, where we have used the definition eqn (1.6) of ω_{pe}. Thus, by solving the Vlasov equation and Poisson's equation self-consistently for a cold plasma, we have obtained the familiar expression for the frequency of electrostatic waves.

Although plasma kinetic theory must agree with cold plasma theory in the appropriate limit, as in eqn (5.26), its primary purpose is to include finite-temperature effects. We therefore consider an equilibrium electron velocity distribution that is Maxwellian,

$$f_0(v_x,v_y,v_z)=\frac{n_0}{\pi^{\frac{3}{2}}v_{\mathrm{Te}}^3}\exp\{-(v_x^2+v_y^2+v_z^2)/v_{\mathrm{Te}}^2\}, \tag{5.27}$$

where the normalization satisfies eqn (5.16). Let us substitute eqn (5.27) into the kinetic expression that governs electrostatic waves, eqn (5.23). First, we integrate over v_x and v_y, using the identity

$$\int_{-\infty}^{\infty}\exp(-\alpha x^2)\,\mathrm{d}x=(\pi/\alpha)^{\frac{1}{2}}. \tag{5.28}$$

Then, also using eqn (1.6), we obtain

$$1-\frac{2\omega_{\mathrm{pe}}^2}{\pi^{\frac{1}{2}}kv_{\mathrm{Te}}^2}\int_{-\infty}^{\infty}\frac{v_z}{\omega-kv_z}\exp(-v_z^2/v_{\mathrm{Te}}^2)\frac{\mathrm{d}v_z}{v_{\mathrm{Te}}}=0. \tag{5.29}$$

Now the integrand in eqn (5.29) includes a denominator $\omega - kv_z$ which vanishes when

$$v_z = \omega/k. \tag{5.30}$$

Clearly, eqn (5.30) is the condition for the component of an electron's velocity in the direction of wave propagation to be equal to the phase velocity of the wave. Equivalently, the Doppler-shifted frequency $\omega - kv_z$ of the wave in the rest frame of the electron is zero. Thus eqn (5.30) describes a situation in which the phase of the wave that is experienced by the electron does not change. This wave–particle resonance requires the electron to have non-zero velocity, and therefore cannot arise in a cold plasma. The resonance leads to energy transfer by the Landau mechanism, which we shall examine in the next section. Meanwhile, we shall restrict our attention to the case where the value of v_z that satisfies eqn (5.30) greatly exceeds v_{Te}, so that the number of resonant electrons is exponentially small and their effect correspondingly weak. That is

$$\omega/k \gg v_{\mathrm{Te}}. \tag{5.31}$$

We already know that $\omega \simeq \omega_{\mathrm{pe}}$, and since $\lambda_{\mathrm{D}} = v_{\mathrm{Te}}/\omega_{\mathrm{pe}}$ by eqn (1.13), eqn (5.31) is identical to the condition that we met at eqn (1.19), which arose from different physical considerations, namely phase-sampling and screening.

Returning to eqn (5.29) subject to the restriction eqn (5.31), we note that in the range of values of v_z of interest, $kv_z/\omega \ll 1$. We may therefore expand the denominator using the binomial theorem:

$$\frac{1}{\omega - kv_z} = \frac{1}{\omega}\left(1 + \frac{kv_z}{\omega} + \frac{k^2v_z^2}{\omega^2} + \frac{k^3v_z^3}{\omega^3} + \cdots\right). \tag{5.32}$$

Substituting eqn (5.32) into eqn (5.29), we obtain

$$1 - \frac{2\omega_{\mathrm{pe}}^2}{\pi^{\frac{1}{2}}\omega k v_{\mathrm{Te}}^2}\int_{-\infty}^{\infty} v_z\left(1 + \frac{kv_z}{\omega} + \frac{k^2v_z^2}{\omega^2} + \frac{k^3v_z^3}{\omega^3} + \cdots\right)\exp(-v_z^2/v_{\mathrm{Te}}^2)\frac{\mathrm{d}v_z}{v_{\mathrm{Te}}} = 0. \tag{5.33}$$

Now for all positive integers n,

$$\int_{-\infty}^{\infty} x^{2n}\exp(-\alpha x^2)\,\mathrm{d}x = \frac{(2n-1)!}{(2\alpha)^n}(\pi/\alpha)^{\frac{1}{2}}, \tag{5.34}$$

$$\int_{-\infty}^{\infty} x^{2n+1}\exp(-\alpha x^2)\,\mathrm{d}x = 0. \tag{5.35}$$

Using these identities, eqn (5.33) yields

$$1 - \frac{\omega_{\mathrm{pe}}^2}{\omega^2}\left(1 + 3\frac{k^2v_{\mathrm{Te}}^2}{\omega^2} + \cdots\right) = 0. \tag{5.36}$$

The condition eqn (5.31) implies that $3k^2 v_{\mathrm{Te}}^2/\omega^2 \ll 1$ in the range of values of ω and k for which eqn (5.36) is valid. Thus eqn (5.36) indicates that $\omega \simeq \omega_{\mathrm{pe}}$. Accordingly, we replace ω by ω_{pe} in the small correction term. Defining the Debye wavenumber

$$k_{\mathrm{D}} = \omega_{\mathrm{pe}}/v_{\mathrm{Te}}, \tag{5.37}$$

following eqn (1.13), eqn (5.36) gives

$$\omega^2 = \omega_{\mathrm{pe}}^2(1 + 3k^2/k_{\mathrm{D}}^2), \qquad k < k_{\mathrm{D}}. \tag{5.38}$$

This completes the derivation of the thermal correction to the frequency of long-wavelength electrostatic waves that we quoted at eqn (1.21).

Let us now examine the physical significance of the combination of terms that appears in eqn (5.23). A plasma can be considered as a dielectric medium, so that electrostatic waves within the plasma represent an oscillating local polarization. This approach is a little different from our usual one, where plasma properties have been deduced from the combined effects of particles moving in a vacuum, which itself has no dielectric properties. As a result, we have previously been able to write the displacement current simply as $\varepsilon_0 \partial \boldsymbol{E}/\partial t$, as in eqn (1.23). In a dielectric medium, the electric displacement $\boldsymbol{D}$, the polarization $\boldsymbol{P}$, and the relative dielectric permittivity ε are related by

$$\boldsymbol{D} = \varepsilon_0 \boldsymbol{E} + \boldsymbol{P} = \varepsilon\varepsilon_0 \boldsymbol{E}. \tag{5.39}$$

Taking the divergence of eqn (5.39), it follows that

$$\varepsilon = 1 + \frac{\boldsymbol{\nabla} \cdot \boldsymbol{P}}{\varepsilon_0 \boldsymbol{\nabla} \cdot \boldsymbol{E}}. \tag{5.40}$$

The polarization $\boldsymbol{P}$ is produced by the response of the plasma to the electric field $\boldsymbol{E}$. It is related to the polarization charge density ρ_{p} by the definition

$$\boldsymbol{\nabla} \cdot \boldsymbol{P} = -\rho_{\mathrm{p}}. \tag{5.41}$$

The plasma has no net electric charge at equilibrium. Thus the entire charge density associated with the waves is a polarization charge density. Using eqn (5.21), we have

$$\rho_{\mathrm{p}\boldsymbol{k}\omega} = -e\int f_{1\boldsymbol{k}\omega}\,\mathrm{d}^3\boldsymbol{v} = -\frac{\mathrm{i}e^2}{m}E\int \frac{\partial f_0/\partial v_z}{\omega - kv_z}\,\mathrm{d}v_x\,\mathrm{d}v_y\,\mathrm{d}v_z. \tag{5.42}$$

Here, as in eqn (5.23), we have used the fact that, for electrostatic waves, $\boldsymbol{E}$ and $\boldsymbol{k}$ are parallel, and have used their direction to define the z-axis. We can now use eqns (5.41) and (5.42) to give a kinetic expression for $\boldsymbol{\nabla} \cdot \boldsymbol{P}$ in eqn (5.40). The denominator in eqn (5.40) is $\varepsilon_0 \boldsymbol{\nabla} \cdot \boldsymbol{E} = \mathrm{i}\varepsilon_0 kE$, so that we have

$$\varepsilon(k, \omega) = 1 + \frac{e^2}{\varepsilon_0 mk}\int \frac{\partial f_0/\partial v_z}{\omega - kv_z}\,\mathrm{d}v_x\,\mathrm{d}v_y\,\mathrm{d}v_z. \tag{5.43}$$

This is the kinetic expression for the relative dielectric permittivity of a plasma, for an electrostatic wave with frequency ω and wavenumber k. Comparing eqn (5.43) with eqn (5.23), we see that electrostatic normal modes satisfy

$$\varepsilon(k, \omega) = 0. \tag{5.44}$$

Physically, eqns (5.44) and (5.39) state that, although the normal modes involve non-zero $\boldsymbol{E}$, they produce no overall electric displacement $\boldsymbol{D}$ in the plasma. This is equivalent to the result that we obtained at eqn (1.23) in our earliest discussion of electrostatic plasma waves. To see this, let us transform our description back from a dielectric medium to an assembly of charged particles in a vacuum. Let $\boldsymbol{x}$ denote the displacement of electrons in response to the electric field $\boldsymbol{E}$ that produces the local polarization. Then

$$\boldsymbol{P} = -n_0 e\boldsymbol{x} \Rightarrow \frac{\partial \boldsymbol{P}}{\partial t} = -n_0 e\boldsymbol{v} = \boldsymbol{J}. \tag{5.45}$$

Taking the partial derivative of eqn (5.39) with respect to time, using eqn (5.45), and setting $\varepsilon = 0$ as in eqn (5.44), we obtain

$$\varepsilon_0 \frac{\partial \boldsymbol{E}}{\partial t} + \boldsymbol{J} = 0, \tag{5.46}$$

in agreement with eqn (1.23). Thus the kinetic approach is a powerful extension of plasma theory which is fully consistent with our previous results.

5.3 Landau damping

In the preceding section, we noted at eqn (5.30) the possibility of resonance between the wave phase velocity and the velocity of individual electrons. When the resonance condition is satisfied, the denominator in the integrand in eqns (5.23), (5.29), and (5.43) becomes zero. So far, we have been able to neglect the consequences of this resonance by restricting attention to waves that satisfy the inequality eqn (5.31) in a Maxwellian plasma. In general, however, this resonance cannot be neglected. It involves coupling between single-particle and collective aspects of plasma behaviour, and gives rise to an energy flow which is known as Landau damping. Before continuing, we should note that this topic is related to one of the main unsolved questions of physics. It has not yet been possible to resolve fully the contrast between the reversibility in time of microscopic phenomena—for example, the dynamics of a particle described by Newton's laws of motion—and the irreversibility in time of macroscopic phenomena, as described by the second law of the thermodynamics. Any thermodynamic system is in fact constructed from a large number of particles, all of which obey Newton's laws, so that this contrast is central to physics. A resolution of this contrast would be particularly helpful to a full understanding of Landau damping; this is because Landau damping involves a flow of energy between single particles on one hand, and collective excitations of the plasma on the other.

For simplicity, let us suppress two spatial dimensions, so that we have effectively a one-dimensional plasma: a line of electrons, with an equal number of motionless ions. Let us add energy to this plasma in the form of an electrostatic wave. The electrons oscillate back and forth, and an electric field wave propagates along the line with angular frequency ω, wavenumber k, and phase velocity $v_{\rm ph} = \omega/k$. Now let us inject into the plasma a test electron with velocity v. As this electron travels down the line, it will be subject to the force arising from the local electric field. In our model, this field includes a wave component arising from the oscillation of the electrons. If it happens that the test electron is travelling at the same speed as the wave, satisfying the resonance condition $v = v_{\rm ph} = \omega/k$, the wave field that the electron experiences is constant; see Fig. 5.1. Depending on its sign, this constant field will act on the electron to increase or decrease its velocity. Energy will flow between the collective wave excitation of the plasma and the single test electron. The energy flow leads naturally to its own eventual termination: this occurs when the electron velocity has either increased or decreased so much that it differs significantly from the phase velocity of the wave, and is no longer in resonance.

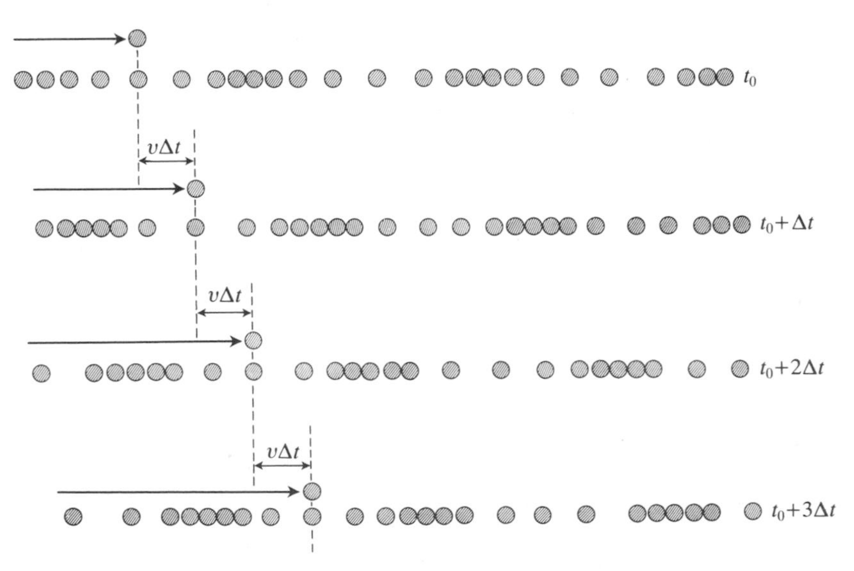

Fig. 5.1 Test electron moving in resonance with a wave field.

There are a number of points to note. First, there is no energy loss; what occurs is a rearrangement of energy between the different aspects of the plasma, collective and single-particle. The sum of the energies of all the electrons in the plasma, including the test electron, remains the same throughout the process. Second, it is clear that the process we have described is a form of collective Coulomb scattering. The electric field experienced by the test electron has its source in the other electrons and the stationary ions. It is the local resultant of all the Coulomb forces which determines whether the velocity of the test electron increases or decreases. Third, we may think of a coherent oscillation as a third type of body with which a test electron may collide, distinct from random electron–electron and electron–ion collisions. For this reason, electrostatic plasma waves are sometimes referred to as plasmons, by analogy with photons and phonons. Fourth, this energy transfer process gives a basic constraint on electrostatic wave propagation in plasmas, in addition to those outlined in Section 1.3. Consider a plasma whose electron velocity distribution is Maxwellian, characterized by a thermal velocity v_{Te}. The number of electrons whose velocity exceeds a few times v_{Te} is exponentially small. It follows that any electrostatic wave whose phase velocity exceeds a few times v_{Te} will meet very few resonant electrons to which it could transfer energy. Conversely, for $v_{\mathrm{ph}} \lesssim v_{\mathrm{Te}}$, such energy transfer may be possible. We conclude that in a Maxwellian plasma, an electrostatic wave will propagate without significant loss of energy provided

$$v_{\mathrm{ph}} \gg v_{\mathrm{Te}}. \tag{5.47}$$

This is the condition that we met previously at eqn (5.31).

So far, we have considered a single test electron with velocity v, and a single electrostatic wave with phase velocity v_{ph} which is supported by the rest of the plasma. We have discussed how, when the resonance condition $v = v_{\mathrm{ph}}$ is satisfied, collective Coulomb scattering causes energy to flow between the field energy of the wave and the kinetic energy of the test electron. This flow occurs because the resonant test electron experiences an effectively constant wave field. Conversely, when v is not close in value to v_{ph}, the test electron experiences an oscillatory wave field. Averaging over many oscillation periods, there is no net effect on the electron. Now, the electrons which compose a real, warm plasma will have a distribution of different velocities v. Thus each individual electron can be regarded as a test electron, moving with its particular velocity through the rest of the plasma. Also, the plasma will support many different electrostatic waves, which have different phase velocities $v_{\mathrm{ph}} = \omega/k$. Clearly, while a few electrons may be in resonance with a given wave, most of the electrons will not be in resonance with most of the waves. In order to calculate the magnitude and direction of the many energy flows, we need a mathematical approach which can quantify the principles that we have discussed. In particular, we need to be able to take into account the number of electrons which have a given velocity. This motivates the application of kinetic theory to this topic. Before doing so, however, let us continue a little further at the level of single-particle dynamics.

Consider test electrons whose velocity v is close to v_{ph} for a given wave, but not exactly equal to it:

$$v = v_{ph}(1+\eta), \qquad |\eta| \ll 1. \tag{5.48}$$

We take an electron which starts at $x = x_0$. The wave phase that this electron experiences is given by

$$\phi(t) = kx(t) - \omega t = kx_0 + \eta k v_{ph} t. \tag{5.49}$$

Here we have used the facts that $x(t) = x_0 + vt$ and $kv_{ph} = \omega$, and then eqn (5.48). Let us write $\phi_0 = kx_0$, a constant. Then, using $v_{ph} = \omega/k$, eqn (5.49) becomes

$$\phi(t) = \phi_0 + \eta\omega t. \tag{5.50}$$

When $\eta = 0$, the apparent phase is exactly constant, $\phi(t) = \phi_0$. When $\eta \neq 0$, with $|\eta| \ll 1$, the apparent phase increases or decreases slowly from $\phi(0) = \phi_0$ as time goes on, depending on the sign of η. Significant changes of phase occur only over the long timescale which follows from eqn (5.50):

$$\tau = \frac{1}{\eta\omega}. \tag{5.51}$$

It follows that over times $t \lesssim \tau$, test electrons with velocity given by eqn (5.48) will experience a wave field that is nearly constant. Thus, a steady increase or decrease of test electron velocity can be expected when v is close to v_{ph}, as well as when v is exactly equal to v_{ph}.

The equation of motion for the electrons is

$$m\frac{\mathrm{d}v}{\mathrm{d}t} = -eE_0 \sin\phi(t), \tag{5.52}$$

for a sine-wave electrostatic field of amplitude E_0. For the case described by eqn (5.48), we use eqn (5.50) for $\phi(t)$ in eqn (5.52) to give

$$m\frac{\mathrm{d}v}{\mathrm{d}t} = -eE_0 \sin(\phi_0 + \eta\omega t). \tag{5.53}$$

We note first of all that the initial phase ϕ_0 is crucial. It is ϕ_0 which determines the sign of $\mathrm{d}v/\mathrm{d}t$ when $0 \lesssim t < \tau$. This is independent of the sign of η, and thus independent of whether the test electron is travelling faster or slower than the wave. The distribution of initial phases is in general random, and when many electrons are present, all values of ϕ_0 between 0 and 2π will be equally probable. To see how a net force on the electron plasma may still arise, we need to consider the implications of the deviation of $\phi(t)$ from ϕ_0 as time passes, when η is finite. Suppose that we have equal numbers of electrons with initial phases ψ_0 and $-\psi_0$. Suppose further that in each group of electrons, n_+ electrons have $v = v_{\mathrm{ph}}(1+\alpha)$, and n_- electrons have $v = v_{\mathrm{ph}}(1-\alpha)$. Then we can relate η in eqn (5.48) to $\pm\alpha$ and, by eqn (5.53), the total force on the group of electrons with initial phase ψ_0 is

$$\begin{aligned} F(\psi_0) &= n_+ m\left(\frac{\mathrm{d}v}{\mathrm{d}t}\right)_{\eta=\alpha,\,\phi_0=\psi_0} + n_- m\left(\frac{\mathrm{d}v}{\mathrm{d}t}\right)_{\eta=-\alpha,\,\phi_0=\psi_0} \\ &= -eE_0\{n_+ \sin(\psi_0 + \alpha\omega t) + n_- \sin(\psi_0 - \alpha\omega t)\}. \end{aligned} \tag{5.54}$$

Similarly, the total force on the group of electrons with initial phase $-\psi_0$ is

$$\begin{aligned} F(-\psi_0) &= n_+ m\left(\frac{\mathrm{d}v}{\mathrm{d}t}\right)_{\eta=\alpha,\,\phi_0=-\psi_0} + n_- m\left(\frac{\mathrm{d}v}{\mathrm{d}t}\right)_{\eta=-\alpha,\,\phi_0=-\psi_0} \\ &= -eE_0\{n_+ \sin(-\psi_0 + \alpha\omega t) + n_- \sin(-\psi_0 - \alpha\omega t)\}. \end{aligned} \tag{5.55}$$

The combined force on all the electrons taken together is

$$\begin{aligned} F &= F(\psi_0) + F(-\psi_0) \\ &= -eE_0\, 2(n_+ - n_-)\cos\psi_0 \sin\alpha\omega t. \end{aligned} \tag{5.56}$$

So there is no net force at $t = 0$, because the forces on the electrons with initial phase ψ_0 cancel the opposite forces on the equal number of electrons with initial phase $-\psi_0$. This is the case, even though the number $2n_+$ of electrons with $v = v_{\mathrm{ph}}(1+\alpha)$ is different from the number $2n_-$ of electrons with $v = v_{\mathrm{ph}}(1-\alpha)$. Only as t increases from zero does the effect of having different numbers of electrons with different phase evolution become apparent, so that a net force arises; see Fig. 5.2. It is this force which transfers energy between the electrons and the wave field.

In conclusion, eqns (5.53) and (5.56) tell us the following about the energy flow. First, the initial phase ϕ_0 determines whether a given resonant electron increases or decreases its velocity in the wave field. Second, a net force can occur in an assembly of electrons in which equal numbers have opposite initial phases, provided that there is a gradient in the velocity distribution near $v = v_{\mathrm{ph}}$: in our simple case, $n_+ \neq n_-$. This force has its origin in the different phase evolution of electrons with different velocities. Third, the magnitude and direction of this force depends on the gradient of the velocity distribution at $v = v_{\mathrm{ph}}$—in our case, on $n_+ - n_-$—and not on the value of the velocity distribution at $v = v_{\mathrm{ph}}$. We have arrived at these conclusions by using single-particle dynamics. Let us now develop the subject further, using kinetic theory to give a full description of the process which is known as Landau damping.

We begin by returning to eqn (5.43). As we have seen, this kinetic expression for the dielectric response of the plasma possesses three relevant features. First, the integrand includes the possibility of wave-

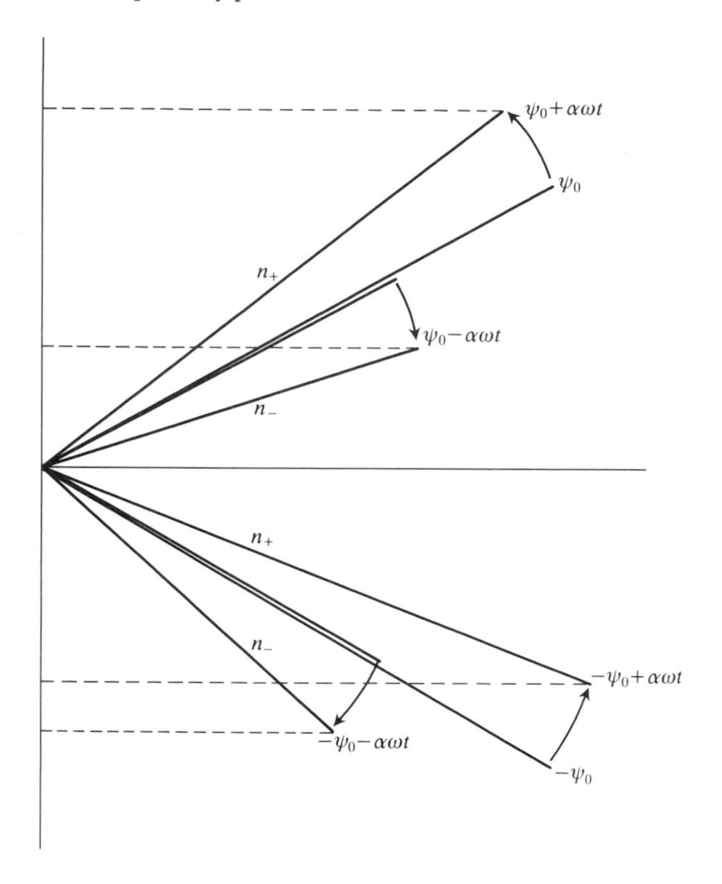

Fig. 5.2 Phase evolution of electrons.

particle resonance. Second, the integrand also depends on the velocity-space gradient $\partial f_0/\partial v_z$ in the direction of propagation of the wave; we have seen in eqn (5.56) that this gradient enters our expression for the force on resonant electrons. Third, we saw at eqn (5.36) that eqn (5.23)—or equivalently eqn (5.44)—yields the dispersion relation for electrostatic waves. Thus, from what we already know about the properties of $\varepsilon(k, \omega)$, it appears that eqn (5.43) contains the information that we require for a kinetic treatment of Landau damping.

To extract this information, we require a result from complex analysis. When the integration over v_z is carried out in eqn (5.43), the contribution to $\varepsilon(k, \omega)$ from $v_z = \omega/k$ is the imaginary term

$$\mathrm{i}\varepsilon_i(k, \omega) = -\frac{\mathrm{i}\pi e^2}{\varepsilon_0 m k^2}\int\left(\frac{\partial f_0}{\partial v_z}\right)_{v_z=\omega/k} \mathrm{d}v_x\, \mathrm{d}v_y. \tag{5.57}$$

This contribution is additional to the real part $\varepsilon_r(k, \omega)$ which corresponds to the expressions such as eqn (5.36). Thus, in general, the normal modes will be described by

$$\varepsilon(k, \omega) = \varepsilon_r(k, \omega) + \mathrm{i}\varepsilon_i(k, \omega) = 0, \tag{5.58}$$

following eqn (5.44). It is usual to assume that

$$|\varepsilon_i| \ll |\varepsilon_r|, \tag{5.59}$$

so that the characteristics of the wave are still primarily determined by ε_r. We aim to solve eqn (5.58) for ω, given k. It is clear that ω will have both a real part and a small imaginary part; let us write

$$\omega = \omega_r - \mathrm{i}\gamma, \qquad |\gamma| \ll 1. \tag{5.60}$$

The time-dependence then becomes

$$\exp(-\mathrm{i}\omega t) = \exp(-\mathrm{i}\omega_r t)\exp(-\gamma t), \tag{5.61}$$

so that a positive value of γ corresponds to damping of a wave that has frequency ω_r. Now consider a Taylor expansion of eqn (5.58), using eqn (5.60) and the inequality eqn (5.59):

$$\varepsilon(k, \omega) = \varepsilon_r(k, \omega_r) + \mathrm{i}\varepsilon_i(k, \omega_r) - \mathrm{i}\gamma\left(\frac{\partial \varepsilon_r}{\partial \omega}(k, \omega)\right)_{\omega=\omega_r} + \ldots = 0. \tag{5.62}$$

Let us separate real and imaginary parts in eqn (5.62). It follows that the real frequency ω_r is a solution of

$$\varepsilon_r(k, \omega_r) = 0, \tag{5.63}$$

and that

$$\gamma = \frac{\varepsilon_i(k, \omega_\mathrm{r})}{\left(\frac{\partial \varepsilon_r}{\partial \omega}(k, \omega)\right)_{\omega=\omega_r}}. \tag{5.64}$$

Now eqn (5.64) is a general result, which applies to any relation of the form given by eqn (5.58). For electrostatic waves in a plasma, ε_i is given by eqn (5.57), so that eqn (5.64) becomes

$$\gamma = \frac{-\pi e^2}{\varepsilon_0 m k^2 \left(\frac{\partial \varepsilon_r}{\partial \omega}(k, \omega)\right)_{\omega=\omega_r}} \int \left(\frac{\partial f_0}{\partial v_z}\right)_{v_z=\omega/k} \mathrm{d}v_x\, \mathrm{d}v_y. \tag{5.65}$$

As we expected from eqn (5.56), the velocity gradient of f_0 at $v_z = \omega/k$ determines both the sign and magnitude of γ, and hence of the energy flow.

We can now calculate the Landau damping rate γ for electrostatic waves in a Maxwellian plasma, where f_0 is given by eqn (5.27), so that

$$\left(\frac{\partial f_0}{\partial v_z}\right)_{v_z=\omega/k} = -\frac{2\omega}{k v_{\mathrm{Te}}^2}\exp\{-(\omega/k v_{\mathrm{Te}})^2\}\frac{n_0}{\pi^{\frac{3}{2}} v_{\mathrm{Te}}^3}\exp\{-(v_x^2 + v_y^2)/v_{\mathrm{Te}}^2\}. \tag{5.66}$$

Substituting eqn (5.66) into eqn (5.57), we carry out the integration over v_x and v_y using eqns (5.28) and (1.6) to obtain

$$\mathrm{i}\varepsilon_i(k, \omega_r) = 2\mathrm{i}\pi^{\frac{1}{2}} \frac{\omega_r \omega_{\mathrm{pe}}^2}{k^3 v_{\mathrm{Te}}^3} \exp\{-(\omega_r / k v_{\mathrm{Te}})^2\}. \tag{5.67}$$

We shall use this expression in the numerator of eqn (5.64). Turning to the denominator, we already know from eqn (5.36) that

$$\varepsilon_r(k, \omega) = 1 - \frac{\omega_{\mathrm{pe}}^2}{\omega^2}\left(1 + \frac{3k^2 v_{\mathrm{Te}}^2}{\omega^2}\right). \tag{5.68}$$

From eqns (5.63) and (5.68), we know that to leading order $\omega_r = \omega_{\mathrm{pe}}$, so that differentiating eqn (5.68) with respect to ω yields, to good approximation,

$$\left(\frac{\partial \varepsilon_r}{\partial \omega}(k, \omega)\right)_{\omega = \omega_r} = \frac{2}{\omega_{\mathrm{pe}}}. \tag{5.69}$$

Substituting eqns (5.67) and (5.69) into eqn (5.64), we obtain

$$\gamma = \omega_{\mathrm{pe}} \pi^{\frac{1}{2}} \frac{k_{\mathrm{D}}^3}{k^3} \exp\{-(k/k_{\mathrm{D}})^2\}. \tag{5.70}$$

Thus, for a Maxwellian plasma, γ is a positive quantity, so that eqn (5.70) gives the rate at which electrostatic waves undergo Landau damping. Using kinetic theory, we have extended the single-particle approach of the earlier part of this section, and have integrated the contribution to Landau damping of all electrons in the plasma.

5.4 The Fokker–Planck equation and binary Coulomb collisions

So far, our discussion of plasma kinetic theory has not included the effects of binary Coulomb collisions. The electric and magnetic fields that we considered in the Vlasov equation originated from the collective plasma charge density and current density, as described at eqns (5.14) and (5.15). Microscopic fields associated with close approaches between pairs of particles were excluded. In this section, we shall see how the effects of Coulomb collisions can be included within the framework of kinetic theory. This involves the combination of collective and single-particle aspects of plasma behaviour. Let us start by recalling two basic ideas associated with the concepts of Debye length (Section 1.2) and of binary Coulomb collisions (Section 1.4). First, the electric field associated with a given particle is effectively screened from the rest of the plasma for distances that exceed λ_{D} given by eqn (1.13), provided that the number N_{D} of particles within the Debye sphere given by eqn (1.18) greatly exceeds unity. Second, the long-range nature of the Coulomb force leads to the expression eqn (1.27) for the scattering angle; this tells us that most charged particles in the plasma interact simultaneously, and with comparable strength, with a number of scattering centres. These two conclusions led us naturally to a study of the collective properties of the plasma. Now, however, we shall examine a corollary of these points: some of the particle interactions in the plasma cannot be described in terms of collective, averaged fields. If a given particle A penetrates deep inside the Debye sphere of particle B, it will interact with the unscreened electric field of B. This interaction will be essentially two-particle, not collective. Equivalently, the impact parameter b of A with respect to B may be so small that the Coulomb scattering deflection ϕ given by eqn (1.27) is large. This is strong Coulomb scattering from a single centre, not simultaneous weaker scattering from a number of centres. We refer to these two-particle non-collective interactions as 'collisions'.

The effect of each collision is to change the magnitude and direction of the velocities of the particles rather sharply on a very short timescale. Viewed on a longer timescale, these changes appear nearly discontinuous. Suppose that we follow a given particle through a sequence of many collisions. We can plot the evolution in time of the magnitude of its velocity as in Fig. 5.3. Or we can plot the evolution of the x and y components of velocity using vectors as in Fig. 5.4. Each sharp change in Fig. 5.3 corresponds to a velocity change arrow in Fig. 5.4.

It is clear that we need to deal with plasma particles, each of which is displaying two kinds of superimposed motion in velocity space. These are a smooth, continuous evolution in response to the local average fields; and a random walk under the influence of two-particle non-collective Coulomb interactions. There are two distinct stages in representing this second process mathematically. First, there is the general question of an ensemble of particles which are performing a random walk in velocity space in response to sequential quasi-impulsive forces. This topic has

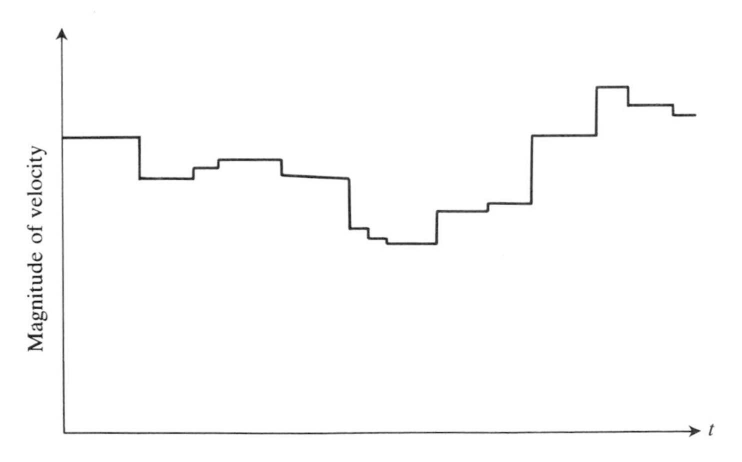

Fig. 5.3 Time evolution of the magnitude of velocity.

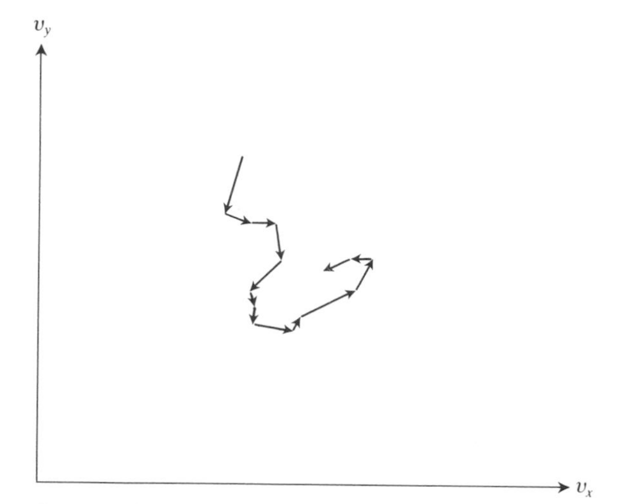

Fig. 5.4 Evolution of x and y velocity components.

wide applications in physics and beyond—it is by no means specific to plasma physics. For our purposes, it is dealt with by the Fokker–Planck equation, which we shall discuss later in this Section. The second stage concerns the Coulomb interaction. In a plasma, it is the Coulomb force which determines the nature of the two-particle interactions whose effects are described by the Fokker–Planck equation. It follows that we shall need to study the Coulomb interaction in greater detail in order to apply the Fokker–Planck equation to collisions in plasmas.

The process of collisional relaxation of a velocity distribution is fairly familiar, and we expect the Fokker–Planck equation to reproduce its features. Suppose that we have a background Maxwellian plasma of temperature T, and inject into it a much smaller number of beam particles whose velocities all lie close to some value v_0 in a particular direction; see Fig. 5.5. We know that, ultimately, collisional relaxation of this distribution will have the following effect: a group of particles that started out as beam particles will be indistinguishable from a group of particles that started out in the background plasma. Both groups will have the same mean velocity, very close to zero by conservation of momentum, and the same thermal spread, characterized by T. Collisions with the background plasma thus have two distinct effects on the beam particles. First, there is a general loss of directed velocity: the mean velocity of the beam particles starts at v_0, but ends close to zero, as if the beam experienced friction with the background plasma. Second, there is diffusion in velocity space: the beam particles start with velocities which are all very close to each other, and end with a distribution of velocities that has the same thermal spread as the background plasma. We expect

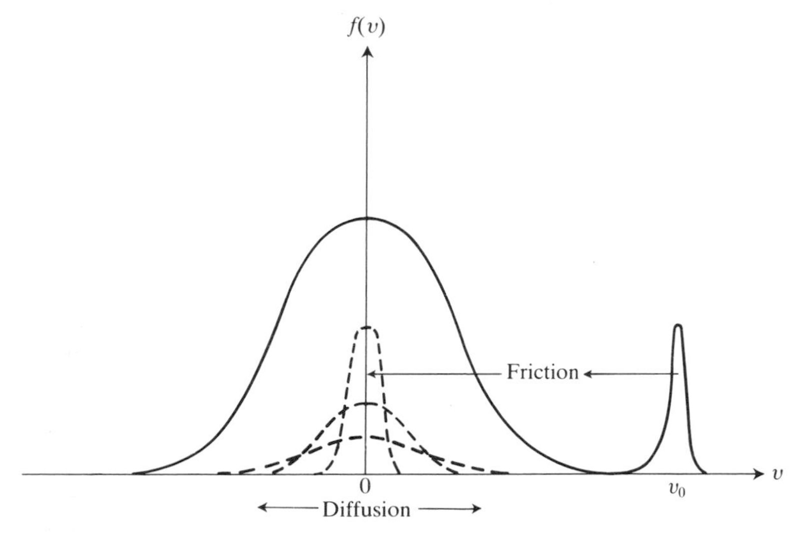

Fig. 5.5 Evolution of beam particles injected into background Maxwellian plasma.

the Fokker–Planck equation to describe both the frictional and the diffusive effects. We note also that there is a contrast between the microscopic reversibility of the Coulomb collisional process, and the macroscopic irreversibility of the relaxation of the velocity distribution. The collisions act as a randomizing process by which the entropy of the system can be increased, while its structure—in the sense of deviation from the Maxwellian—is eliminated.

Now let us construct the Fokker–Planck equation. It is essentially probabilistic. The key quantity is the probability $P(\boldsymbol{v}, \Delta\boldsymbol{v})$ that during the brief time Δt, the velocity of a particle will change its value from $\boldsymbol{v}$ to $\boldsymbol{v} + \Delta\boldsymbol{v}$ as a result of collisions. Note that $P(\boldsymbol{v}, \Delta\boldsymbol{v})$ is independent of the time at which the collision occurs, and is determined in our case by the properties of the Coulomb interaction. We shall return to this in the second stage of our calculation. The fraction of the total number of particles which have velocity $\boldsymbol{v}$ at time t is $f(\boldsymbol{v}, t)$—any spatial dependence of f is not relevant. It follows that the fraction of the total number of particles which both start with velocity $\boldsymbol{v}$ at time t and end with velocity $\boldsymbol{v} + \Delta\boldsymbol{v}$ at time $t + \Delta t$ is

$$f(\boldsymbol{v}, t)P(\boldsymbol{v}, \Delta\boldsymbol{v}). \tag{5.71}$$

The total number of particles which attain velocity $\boldsymbol{v}$ at time t is determined by the outcome of collisions in the preceding time interval Δt. This number is obtained, following eqn (5.71), by considering terms of the form

$$f(\boldsymbol{v} - \Delta\boldsymbol{v}, t - \Delta t)P(\boldsymbol{v} - \Delta\boldsymbol{v}, \Delta\boldsymbol{v}) \tag{5.72}$$

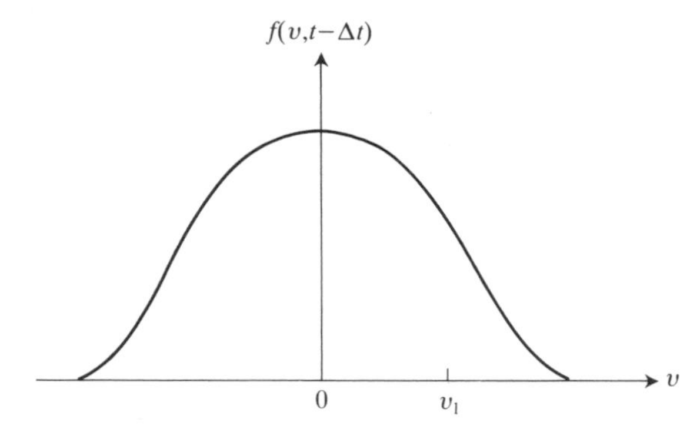

Fig. 5.6 $f(v, t - \Delta t)$.

which gives the fraction of particles which both start with velocity $\boldsymbol{v} - \Delta\boldsymbol{v}$ at time $t - \Delta t$ and end with velocity $\boldsymbol{v}$ at time t. We will need to sum many such terms for a given value of $\boldsymbol{v}$, each term having a different initial velocity $\boldsymbol{v} - \Delta\boldsymbol{v}$. For example, suppose that (in one dimension, for simplicity) $f(v, t - \Delta t)$ has the shape shown in Fig. 5.6. Consider a particular value v_1 of v, and aim to calculate $f(v_1, t)$. First, using Fig. 5.6, $f(v_1 - \Delta v, t - \Delta t)$ has the shape shown in Fig. 5.7. Next, suppose that the collision characteristics for v_1 are given by $P(v_1 - \Delta v, \Delta v)$ shown in Fig. 5.8. Then for each value of Δv, eqn (5.72) with $v = v_1$ corresponds to the product of the heights of the graphs in Figs. 5.7 and 5.8. We require the sum of these products for all values of Δv. Hence in general,

$$f(\boldsymbol{v}, t) = \int f(\boldsymbol{v} - \Delta\boldsymbol{v}, t - \Delta t)P(\boldsymbol{v}, \Delta\boldsymbol{v})\, \mathrm{d}^3\Delta\boldsymbol{v}. \tag{5.73}$$

We now make an assumption: suppose that $P(\boldsymbol{v}, \Delta\boldsymbol{v})$ is such that the integrand in eqn (5.73) is strongly weighted towards terms with $|\Delta\boldsymbol{v}| \ll |\boldsymbol{v}|$. Then we can use a Taylor expansion in $\Delta\boldsymbol{v}$, truncated after two

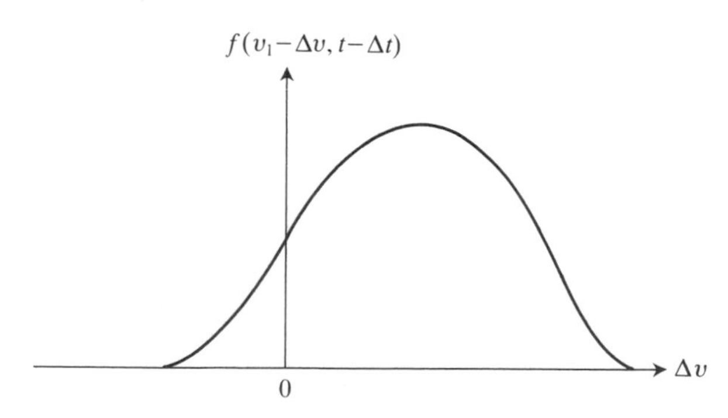

Fig. 5.7 $f(v_1 - \Delta v, t - \Delta t)$.

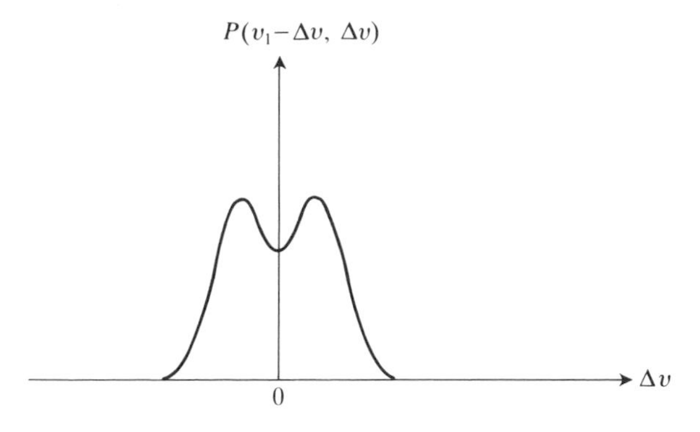

Fig. 5.8 $P(v_1 - \Delta v, \Delta v)$.

terms, to good approximation:

$$
\begin{aligned}
f(\boldsymbol{v} - \Delta\boldsymbol{v}, t - \Delta t)P(\boldsymbol{v} - \Delta\boldsymbol{v}, \Delta\boldsymbol{v}) = {} & f(\boldsymbol{v}, t - \Delta t)P(\boldsymbol{v}, \Delta\boldsymbol{v}) \\
& - \Delta\boldsymbol{v} \cdot \frac{\partial}{\partial \boldsymbol{v}} \{f(\boldsymbol{v}, t - \Delta t)P(\boldsymbol{v}, \Delta\boldsymbol{v})\} \\
& + \frac{1}{2}\sum_{i,j} \Delta v_i \, \Delta v_j \frac{\partial}{\partial v_i}\frac{\partial}{\partial v_j}\{f(\boldsymbol{v}, t - \Delta t)P(\boldsymbol{v}, \Delta\boldsymbol{v})\}.
\end{aligned}
\tag{5.74}
$$

Now Taylor expand in t, requiring that Δt is small:

$$
f(\boldsymbol{v}, t - \Delta t) = f(\boldsymbol{v}, t) - \Delta t \left.\frac{\partial}{\partial t}\right)_c f(\boldsymbol{v}, t) + \ldots . \tag{5.75}
$$

Here $\partial/\partial t)_c$ denotes the contribution to $\partial/\partial t$ that arises from collisions, whose effect is contained in $P(\boldsymbol{v}, \Delta\boldsymbol{v})$. Also, note that $\Delta\boldsymbol{v}$ is independent of $\boldsymbol{v}$: when we operate with $\partial/\partial\boldsymbol{v}$ on $\Delta\boldsymbol{v}$, $\Delta\boldsymbol{v}$ is treated as a constant. We now combine eqns (5.74) and (5.75), to obtain

$$
\begin{aligned}
& f(\boldsymbol{v} - \Delta\boldsymbol{v}, t - \Delta t)P(\boldsymbol{v} - \Delta\boldsymbol{v}, \Delta\boldsymbol{v}) \\
& \quad = f(\boldsymbol{v}, t)P(\boldsymbol{v}, \Delta\boldsymbol{v}) - \Delta t \left.\frac{\partial}{\partial t}\right)_c f(\boldsymbol{v}, t)P(\boldsymbol{v}, \Delta\boldsymbol{v}) - \frac{\partial}{\partial \boldsymbol{v}} \cdot \{f(\boldsymbol{v}, t)\Delta\boldsymbol{v}P(\boldsymbol{v}, \Delta\boldsymbol{v})\} \\
& \qquad + \frac{1}{2}\sum_{i,j} \frac{\partial}{\partial v_i}\frac{\partial}{\partial v_j}\{f(\boldsymbol{v}, t)\Delta v_i \, \Delta v_j P(\boldsymbol{v}, \Delta\boldsymbol{v})\},
\end{aligned}
\tag{5.76}
$$

where we have neglected the remaining higher-order terms arising from eqn (5.75). Now $P(\boldsymbol{v}, \Delta\boldsymbol{v})$ is a probability, so that it must satisfy

$$
\int P(\boldsymbol{v}, \Delta\boldsymbol{v}) \, \mathrm{d}^3\Delta\boldsymbol{v} = 1. \tag{5.77}
$$

Using this fact, we substitute eqn (5.76) into eqn (5.73) to obtain

$$\Delta t \left.\frac{\partial}{\partial t}\right)_{\mathrm{c}} f(\boldsymbol{v}, t) = -\frac{\partial}{\partial \boldsymbol{v}} \cdot \left\{ f(\boldsymbol{v}, t) \int \Delta \boldsymbol{v} P(\boldsymbol{v}, \Delta \boldsymbol{v}) \, \mathrm{d}^3 \Delta \boldsymbol{v} \right\}$$
$$+ \frac{1}{2} \sum_{i,j} \frac{\partial}{\partial v_i} \frac{\partial}{\partial v_j} \left\{ f(\boldsymbol{v}, t) \int \Delta v_i \, \Delta v_j P(\boldsymbol{v}, \Delta \boldsymbol{v}) \, \mathrm{d}^3 \Delta \boldsymbol{v} \right\}. \quad (5.78)$$

We now use the shorthand notation

$$\left\langle \frac{\Delta \boldsymbol{v}}{\Delta t} \right\rangle = \frac{1}{\Delta t} \int \Delta \boldsymbol{v} \, P(\boldsymbol{v}, \Delta \boldsymbol{v}) \, \mathrm{d}^3 \Delta \boldsymbol{v}, \quad (5.79)$$

$$\left\langle \frac{\Delta v_i \Delta v_j}{\Delta t} \right\rangle = \frac{1}{\Delta t} \int \Delta v_i \, \Delta v_j P(\boldsymbol{v}, \Delta \boldsymbol{v}) \, \mathrm{d}^3 \Delta \boldsymbol{v}. \quad (5.80)$$

The choice of notation at this stage is always a little problematic. In particular, we note that while $\Delta \boldsymbol{v}$ is a variable independent of $\boldsymbol{v}$, the quantities $\langle \Delta \boldsymbol{v} / \Delta t \rangle$ and $\langle \Delta v_i \Delta v_j / \Delta t \rangle$ defined in eqns (5.79) and (5.80) are both functions of $\boldsymbol{v}$. Physically, $\langle \Delta \boldsymbol{v} / \Delta t \rangle$ is the expected value, for a particle of velocity $\boldsymbol{v}$, of the rate of charge of velocity arising from collisions; $\langle \Delta v_i \Delta v_j / \Delta t \rangle$ is the expected value, for a particle of velocity $\boldsymbol{v}$, of the rate of velocity dispersion arising from collisions. Equivalently, $\langle \Delta \boldsymbol{v} / \Delta t \rangle$ and $\langle \Delta v_i \, \Delta v_j / \Delta t \rangle$ are the expected values per unit time of the changes $\Delta \boldsymbol{v}$ and $\Delta v_i \Delta v_j$ for a particle of velocity $\boldsymbol{v}$. Combining eqns (5.78) to (5.80), we have the Fokker–Planck equation:

$$\left(\frac{\partial}{\partial t}\right)_{\mathrm{c}} f(\boldsymbol{v}, t) = -\frac{\partial}{\partial \boldsymbol{v}} \cdot \left\{ f(\boldsymbol{v}, t) \left\langle \frac{\Delta \boldsymbol{v}}{\Delta t} \right\rangle \right\} + \frac{1}{2} \sum_{i,j} \frac{\partial}{\partial v_i} \frac{\partial}{\partial v_j} \left\{ f(\boldsymbol{v}, t) \left\langle \frac{\Delta v_i \, \Delta v_j}{\Delta t} \right\rangle \right\}. \quad (5.81)$$

On the right-hand side, we have the frictional and diffusive terms that we expected from first principles.

We note that the derivation of the Fokker–Planck equation eqn (5.81) relies on the truncated Taylor series in eqn (5.74). Hence, we can deduce the conditions under which eqn (5.81) is no longer valid. First, if $P(\boldsymbol{v}, \Delta \boldsymbol{v})$ is such that a significant role in eqn (5.73) is played by collisions for which $|\Delta \boldsymbol{v}|$ is not small compared to $|\boldsymbol{v}|$, the truncation after two terms will be inappropriate. Second, even if $P(\boldsymbol{v}, \Delta \boldsymbol{v})$ is such that only collisions involving $|\Delta \boldsymbol{v}| \ll |\boldsymbol{v}|$ matter, a problem may still arise for values of $\boldsymbol{v}$ for which $|\partial f / \partial \boldsymbol{v}|$ is large. In this case, $|\Delta \boldsymbol{v} \cdot \partial f / \partial \boldsymbol{v}|$ may no longer be small compared to f, even though $|\Delta \boldsymbol{v}| \ll |\boldsymbol{v}|$. Again, the truncations in eqns (5.74) and (5.75) will be inappropriate.

We pointed out before eqn (5.71) that the functional form of $P(\boldsymbol{v}, \Delta \boldsymbol{v})$ in a particular physical context depends on the scattering process that $P(\boldsymbol{v}, \Delta \boldsymbol{v})$ represents. In eqns (5.79) and (5.80), it is the functional form of $P(\boldsymbol{v}, \Delta \boldsymbol{v})$ that determines the dependence on $\boldsymbol{v}$ of $\langle \Delta \boldsymbol{v} / \Delta t \rangle$ and $\langle \Delta v_i \, \Delta v_j / \Delta t \rangle$. The Fokker–Planck equation eqn (5.81) is built around these two functions. Thus, in order to use the Fokker–Planck equation in plasma physics, we need to be more specific about the scattering process involved: binary Coulomb collisions.

The only fundamental equation that we require has already been derived. It is eqn (1.27), which relates the scattering angle ϕ to the impact parameter b for a single particle that is incident with velocity v_0 on a scattering centre. In a plasma, however, information is not available about the impact parameter of any given particle incident on a second particle. We can only deal with such a microscopic concept in a probabilistic, averaged way. For example, suppose that we have a broad, uniform stream of test particles, all incident with the same velocity v_0 on a single scattering centre: see Fig. 5.9. Instead of attempting to deal with each incident particle separately, let us aim to calculate an averaged quantity: the mean, taken over all the incident particles, of the square of the scattering angle ϕ. We already know the relation eqn (1.27) between ϕ and the impact parameter b, given v_0. Therefore the first question is, how many incident particles have an impact parameter between b and $b + \mathrm{d}b$? From Fig. 5.9, it is clear that this number is proportional to the quantity

$$\mathrm{d}\sigma = 2\pi b \, \mathrm{d}b, \tag{5.82}$$

which is known as the scattering cross-section. We then define the mean squared scattering angle $\langle \phi^2(v_0) \rangle$ for the stream of particles with velocity v_0 by

$$\langle \phi^2(v_0) \rangle = \frac{\int_{b_{\min}}^{b_{\max}} \phi^2(b, v_0) \, \mathrm{d}\sigma(b)}{\int_{b_{\min}}^{b_{\max}} \mathrm{d}\sigma(b)}, \tag{5.83}$$

where $\phi^2(b, v_0)$ follows from eqn (1.27). Here $b_{\max}$ is the maximum value of the impact parameter for which a treatment using binary Coulomb collisions is appropriate. By our previous comments,

$$b_{\max} = \lambda_{\mathrm{D}} \tag{5.84}$$

where λ_{D} is the Debye length given by eqn (1.13); we know that for

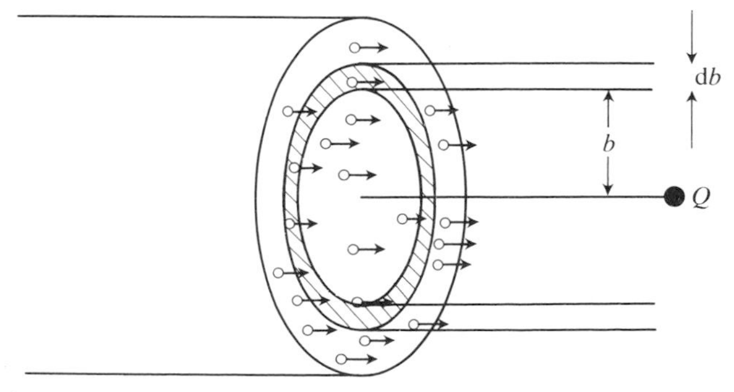

Fig. 5.9 Test particles incident on single scattering centre.

$b > \lambda_D$, a collective treatment applies. The need for a lower bound b_{min} on the range of impact parameters reflects both the simplicity of our model for scattering, and the tendency which we noted in Section 1.4 for integrals involving the long-range Coulomb force to diverge. Our model is reliable so long as the fraction of incident kinetic energy that is converted to electrostatic energy during the approach is not too large. We therefore define

$$b_{min} = qQ/4\pi\varepsilon_0 m v_0^2. \tag{5.85}$$

We note from eqn (5.82) that the integrals in eqn (5.83) are weighted towards the particles which have large impact parameter b and hence, by eqn (1.27), small scattering angle ϕ. We may therefore use, to good approximation, the following expression derived from the small-angle limit of eqn (1.27):

$$\phi^2(b, v_0) = 4\left(\frac{qQ}{4\pi\varepsilon_0 m}\right)^2 \frac{1}{b^2 v_0^4}. \tag{5.86}$$

The integration in eqn (5.83) can now be performed, using eqns (5.82) and (5.86), and the fact that $b_{max} \gg b_{min}$. We obtain

$$\begin{aligned}\langle \phi^2(v_0) \rangle &= 4\left(\frac{qQ}{4\pi\varepsilon_0 m}\right)^2 \frac{1}{v_0^4} \frac{1}{\pi b_{max}^2} \ln(b_{max}/b_{min}) \\ &= \frac{4}{\pi}(b_{min}/b_{max})^2 \ln(b_{max}/b_{min}),\end{aligned} \tag{5.87}$$

where we have substituted for b_{min} using eqn (5.85). Using also eqn (5.82), we have

$$b_{max}/b_{min} = (mv_0^2)/(qQ/4\pi\varepsilon_0\lambda_D). \tag{5.88}$$

It follows that the value of b_{max}/b_{min} is determined physically by the ratio of the kinetic energy of the incident particle to its electrostatic potential energy at a distance of one Debye length from the scattering centre. It is clear from eqn (5.87) that this simple parameter is crucial in determining the averaged effects of binary Coulomb collisions. In a plasma, we are typically concerned with the binary Coulomb collisions of thermal electrons for which $mv_0^2 = k_B T_e$. For this case, b_{max}/b_{min} takes the special value Λ following from eqn (5.88):

$$\Lambda = \frac{k_B T_e 4\pi\varepsilon_0 \lambda_D}{qQ}. \tag{5.89}$$

In scattering theory, the function $\ln(b_{max}/b_{min})$ occurs frequently, as in eqn (5.87). For this reason, there is a further named parameter: the Coulomb logarithm λ_c, defined by

$$\lambda_c = \ln\Lambda = 18 - \ln\left\{\left(\frac{n_e}{10^{19}\ \mathrm{m}^{-3}}\right)^{\frac{1}{2}} \Big/ \left(\frac{k_B T_e}{1\ \mathrm{keV}}\right)^{\frac{3}{2}}\right\}, \tag{5.90}$$

where we have used eqn (1.13) in eqn (5.89). Now we have met the combination $n^{\frac{1}{2}}T^{-\frac{3}{2}}$ before, in the definition eqn (1.18) of the number N_D of particles contained within a Debye sphere. We may write eqn (5.90) in the form

$$\lambda_c = 2 + \ln N_D. \tag{5.91}$$

It is often remarked that $10 < \lambda_c < 20$ for a very wide range of physically realizable plasmas. From eqn (5.91), we see that this is a reflection of the fact that physically realizable plasmas are generally good plasmas, in the sense that they have a very large number of particles within the Debye sphere.

Now that we have introduced the basic quantities involved in Coulomb scattering theory, we can move on to calculate a formula for the quantity $\langle \Delta \boldsymbol{v}/\Delta t \rangle$ given by eqn (5.79). First, let us use eqn (1.27) to write down the relationship between a small change db in the impact parameter and the resulting small change dϕ in scattering angle:

$$\mathrm{d}b = -\frac{qQ}{4\pi\varepsilon_0 m v_0^2}\frac{\mathrm{d}\phi}{2\sin^2(\phi/2)}. \tag{5.92}$$

Now use eqns (1.27) and (5.92) to eliminate b in favour of ϕ in the definition eqn (5.82) of dσ:

$$\mathrm{d}\sigma = 2\pi\left(\frac{qQ}{4\pi\varepsilon_0 m v_0^2}\right)^2 \frac{\mathrm{d}\phi}{2\sin^2(\phi/2)\tan(\phi/2)}, \tag{5.93}$$

It is useful to multiplity the top and bottom of eqn (5.93) by $\sin(\phi/2)$, and use the identity $\sin\phi = 2\sin(\phi/2)\cos(\phi/2)$. This gives

$$\mathrm{d}\sigma = \left(\frac{qQ}{8\pi\varepsilon_0 m v_0^2}\right)^2 \frac{2\pi \sin\phi\, \mathrm{d}\phi}{\sin^4(\phi/2)}. \tag{5.94}$$

Recall from eqn (5.82) that dσ is proportional to the probability that an incident particle has an impact parameter between b and $b + \mathrm{d}b$. We know from eqn (1.27) that to each b, there corresponds a particular scattering angle ϕ. This relation has been used to express dσ in eqn (5.94) as a function of ϕ instead of b. Thus eqn (5.94) expresses dσ as proportional to the probability that a particle is scattered through an angle between ϕ and $\phi + \mathrm{d}\phi$. Let us now use the standard definition of an element of solid angle:

$$\mathrm{d}\Omega = 2\pi \sin\phi\, \mathrm{d}\phi, \tag{5.95}$$

see Fig. 5.10. Substituting eqn (5.95) in eqn (5.94), we obtain

$$\mathrm{d}\sigma = \left(\frac{qQ}{8\pi\varepsilon_0 m v_0^2}\right)^2 \frac{\mathrm{d}\Omega}{\sin^4(\phi/2)} = \mathrm{d}\sigma(v_0, \Omega). \tag{5.96}$$

This is proportional to the probability that a particle is scattered into the

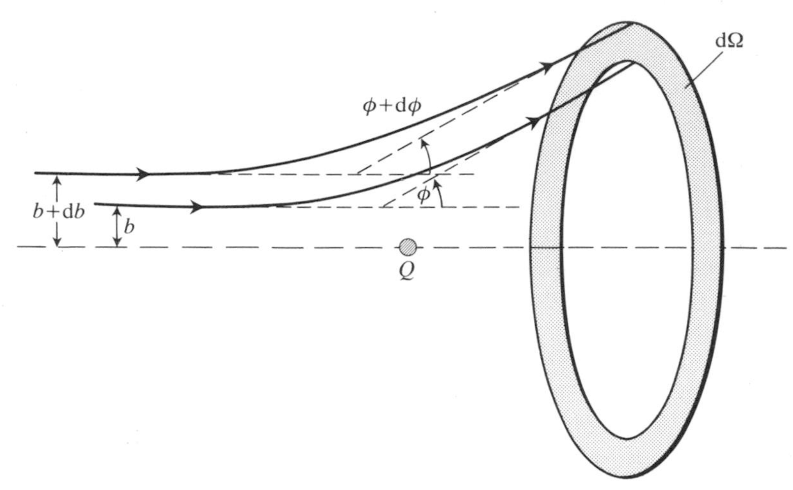

Fig. 5.10 Definition of solid angle.

element of solid angle $d\Omega$, which subtends an angle between ϕ and $\phi + d\phi$. The $\text{cosec}^4(\phi/2)$ dependence is characteristic of the famous Rutherford scattering formula eqn (1.27).

We now wish to calculate the mean change in velocity for the stream of particles incident with velocity v_0 on the scattering centre, depicted in Fig. 5.9. Note first that this system has rotational symmetry about the axis passing down the centre of the stream to the scattering centre. It follows from this symmetry that the sum of all the velocity changes, projected onto the plane perpendicular to this axis, is zero. We therefore need to consider only the change in the component of velocity parallel to this axis. Now the magnitude of the velocity of each incident particle remains constant at v_0 throughout the scattering process. Only the direction changes, by a total angle ϕ. Hence, see Fig. 5.11, the change in the component of velocity parallel to the axis is

$$\Delta v_{\parallel} = -2(v_0 \sin(\phi/2)) \sin(\phi/2) = -2v_0 \sin^2(\phi/2). \tag{5.97}$$

The probability per unit time that a particle moving with velocity v_0 is scattered into the element of solid angle $d\Omega$ is $v_0\, d\sigma(v_0, \Omega)$. The factor of v_0 arises because the probability that an incident particle will reach a given scattering centre within a given period of time is proportional to the speed at which it is moving. We are concerned with probability per unit time because this is the form of probability that is required in the Fokker–Planck equation, as in eqns (5.79) and (5.80). The average change per unit time in the component of velocity parallel to the collision axis is therefore

$$\left\{\frac{\Delta v_{\parallel}}{\Delta t}\right\} = \int_{\Omega} \Delta v_{\parallel} v_0\, d\sigma(v_0, \Omega), \tag{5.98}$$

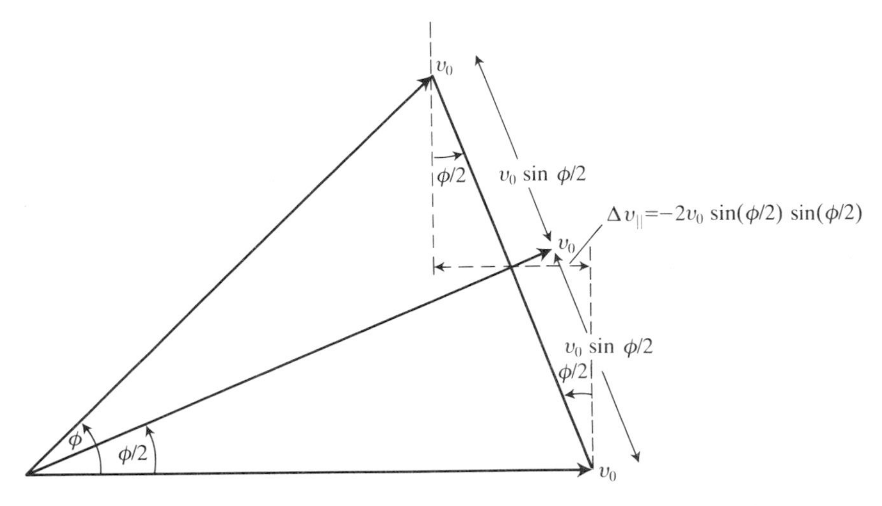

Fig. 5.11 Change of parallel velocity due to scattering.

where the integration is performed over all solid angle. Note that $\{\Delta v_{\|}/\Delta t\}$ remains a function of v_0, just as $\langle \Delta \boldsymbol{v}/\Delta t\rangle$ given by eqn (5.79) is itself a function of $\boldsymbol{v}$. We now carry out the integration over solid angle in eqn (5.98), using eqns (5.94) and (5.97):

$$\left\{\frac{\Delta v_{\|}}{\Delta t}\right\} = -\left(\frac{qQ}{8\pi\varepsilon_0 m v_0^2}\right)^2 4\pi v_0^2 \int \frac{\sin^2(\phi/2)\sin\phi\, \mathrm{d}\phi}{\sin^4(\phi/2)}. \tag{5.99}$$

As in eqn (5.83), there are physical limits on the range of ϕ which it is appropriate to consider in the integration in eqn (5.99). These are $\phi_{\min}$, corresponding to the maximum impact parameter $b_{\max}$ for which a binary treatment is appropriate, given by eqn (5.84); and $\phi_{\max}$, corresponding to the smallest realistic impact parameter $b_{\min}$ given by eqn (5.85). The integral in eqn (5.99) can thus be written, using again $\sin\phi = 2\sin(\phi/2)\cos(\phi/2)$, as

$$\begin{aligned}\int_{\phi_{\min}}^{\phi_{\max}} \frac{\sin^2(\phi/2)\sin\phi\,\mathrm{d}\phi}{\sin^4(\phi/2)} &= \int_{\phi_{\min}}^{\phi_{\max}} 4\frac{\cos(\phi/2)}{\sin(\phi/2)}\mathrm{d}(\phi/2)\\ &= \int_{\phi_{\min}}^{\phi_{\max}} 4\frac{\mathrm{d}\sin(\phi/2)}{\sin(\phi/2)}\\ &= 4\ln(\sin(\phi_{\max}/2)/\sin(\phi_{\min}/2)). \end{aligned} \tag{5.100}$$

If both $\phi_{\max}/2$ and $\phi_{\min}/2$ are sufficiently small that their cosine is approximately unity, we can use eqn (1.27) to write eqn (5.100) as $4\ln(b_{\max}/b_{\min})$. Then eqn (5.99) can be written

$$\left\{\frac{\Delta v_{\|}}{\Delta t}\right\} \simeq -\pi\left(\frac{qQ}{4\pi\varepsilon_0 m v_0}\right)^2 \ln\left(\frac{b_{\max}}{b_{\min}}\right). \tag{5.101}$$

Note that, using the discussion of eqns (5.88) to (5.90), it is possible to express eqn (5.101) in terms of the Coulomb logarithm. We note also that in the Fokker–Planck equation, $\langle \Delta \boldsymbol{v}/\Delta t \rangle$ is a vector quantity. Since $\Delta v_{\parallel}$ is the component of velocity change along a direction whose unit vector is $\boldsymbol{v}_0/v_0$, eqn (5.101) gives

$$\left\{\frac{\Delta \boldsymbol{v}}{\Delta t}\right\} = -\pi\left(\frac{qQ}{4\pi\varepsilon_0 m}\right)^2 \ln\left(\frac{b_{\max}}{b_{\min}}\right)\frac{\boldsymbol{v}_0}{v_0^3}. \tag{5.102}$$

This effectively completes our derivation from first principles of a velocity-dependent coefficient in the Fokker–Planck equation. The quantity $\{\Delta \boldsymbol{v}/\Delta t\}$ differs from $\langle \Delta \boldsymbol{v}/\Delta t \rangle$ by two minor features which are associated with the simplicity of our model for collisions. First, $\langle \Delta \boldsymbol{v}/\Delta t \rangle$ is calculated in the centre of mass frame. Second, in general a large number of scattering centres will need to be considered; their velocities will be non-zero, and described by a distribution function. We shall not pursue these points here, as all the basic concepts of the topic have now been discussed. We note finally that in general, the stream of particles incident on a given scattering centre in a plasma will include particles with many different initial velocities. This stream can be divided up into sub-populations, each incident with a particular velocity $\boldsymbol{v}_0$. The evolution of each sub-population in velocity space will follow the treatment of this section. It will be governed by the Fokker–Planck equation for $f(\boldsymbol{v}_0, t)$, where $\boldsymbol{v}_0$ is the particular value of the initial velocity of the sub-population. This emphasizes the generality of the Fokker–Planck equation.

We began by pointing out that some particles in a plasma will approach each other to distances which are less than a Debye length λ_D. A description of such collisions in terms of collective fields is not possible. The number of such collisions may be very large, and the result of each collision is sensitive to the initial velocities and positions of the two particles involved. This calls for a probabilistic treatment, which develops in two stages. First, the Fokker–Planck equation is derived. This is a general equation which describes the time evolution of an ensemble of particles undergoing quasi-impulsive velocity changes. It involves averaged quantities $\langle \Delta \boldsymbol{v}/\Delta t \rangle$ and $\langle \Delta v_i\, \Delta v_j/\Delta t \rangle$ which are themselves functions of velocity; the form of these functions is determined by the nature of the force which mediates the collisions. This leads us to the second stage: the calculation of such terms for the binary Coulomb collisions that occur in a plasma. The nature of this calculation is determined by the absence of detailed information on particle trajectories, and hence relative impact parameters. It is necessary to integrate over all relevant impact parameters, or equivalently scattering angles, in order to adjust to this lack of information. The resulting expressions are relatively simple. It is noteworthy that they generally have a hybrid character in which collective quantities such as the Debye length combine with single-particle quantities.

5.5 Two-stream instability and negative energy waves

We shall now consider plasmas whose velocity distributions deviate strongly from a Maxwellian. In particular, we consider beam-plasma systems, where the distribution in velocity of one of the plasma species—for example, the electrons—has two distinct components. These are a background, Maxwellian component, and an energetic component whose velocity in a particular direction greatly exceeds that of the background. The energetic component is referred to as a beam, and it represents a source of free energy. In order to examine how this free energy may be released, we must investigate the stability of the beam-plasma system. This requires the calculation of its response to small perturbations, for which we employ the techniques of Section 5.2. For simplicity, we assume that the difference in velocity between the beam electrons and the background electrons greatly exceeds the spread in velocity of either population. That is, we consider a beam-plasma system which, to leading order, is cold. In this approximation, the initial electron velocity distribution function $f_0(\boldsymbol{v})$ is zero everywhere except at $\boldsymbol{v} = 0$ or at $v_z = v_0$, with v_x and v_y negligible, where v_0 is the velocity of the beam, whose direction we choose to define the z-axis. We shall consider electrostatic perturbations, and choose the wavevector $\boldsymbol{k}$ to lie along the z-axis, parallel to the beam. Physically, the suppression of k_x and k_y does not matter. The propagation of electrostatic waves in these directions, for the unmagnetized plasma that we consider, is governed by the distribution of electron velocities in the x- and y-directions. These distributions have no beam component, so that the previous cold plasma (or, to next order, Maxwellian plasma) treatment will apply. Thus, we restrict attention to the only component of $\boldsymbol{k}$ for which the presence of the beam is significant. We denote the fraction of electrons in the beam by ξ, so that the remaining fraction $(1 - \xi)$ forms the background plasma. Using the cold plasma expression for $\varepsilon(k, \omega) = 0$ given by eqn (5.25), we obtain for the cold beam-plasma system

$$\varepsilon(k, \omega) = 1 - (1 - \xi)\frac{\omega_{pe}^2}{\omega^2} - \frac{\xi\omega_{pe}^2}{(\omega - kv_0)^2} = 0, \qquad (5.103)$$

where we have employed eqn (1.6).

In the limit where the number of electrons in the beam tends to zero, ξ vanishes and eqn (5.103) reduces to eqn (5.26). We note also that $(\omega - kv_0)$ is the Doppler-shifted wave frequency that is experienced in the rest frame of the beam electrons. It is clear from eqn (5.103) that we may regard the background and beam populations as distinct plasmas, both with their own plasma frequency proportional to their density. We define

$$\omega_{\mathrm{po}}^2 = (1 - \xi)\omega_{\mathrm{pe}}^2, \tag{5.104}$$

$$\omega_{\mathrm{pb}}^2 = \xi\omega_{\mathrm{pe}}. \tag{5.105}$$

Then eqn (5.103) can be written

$$\varepsilon(k, \omega) = 1 - \frac{\omega_{\mathrm{po}}^2}{\omega^2} - \frac{\omega_{\mathrm{pb}}^2}{(\omega - kv_0)^2} = 0. \tag{5.106}$$

This dispersion relation is approximately satisfied under two conditions. Either

$$\omega^2 \simeq \omega_{\mathrm{po}}^2, \tag{5.107}$$

so that $\omega_{\mathrm{po}}^2/\omega^2 \simeq 1$,cancelling the first term in the expression for $\varepsilon(k, \omega)$; or

$$(\omega - kv_0)^2 \simeq \omega_{\mathrm{pb}}^2, \tag{5.108}$$

so that $\omega_{\mathrm{pb}}^2/(\omega - kv_0)^2 \simeq 1$ and eqn (5.106) is again approximately satisfied. Thus, both the background and the beam components of the plasma support two families of electrostatic waves, given by eqns (5.107) and (5.108) respectively. We note that we may write eqn (5.106) in the form

$$(\omega^2 - \omega_{\mathrm{po}}^2)\{(\omega - kv_0)^2 - \omega_{\mathrm{pb}}^2\} = \omega_{\mathrm{pb}}^2\omega_{\mathrm{po}}^2. \tag{5.109}$$

Using the binomial theorem, this becomes

$$(\omega - \omega_{\mathrm{po}})(\omega + \omega_{\mathrm{po}})(\omega - kv_0 - \omega_{\mathrm{pb}})(\omega - kv_0 + \omega_{\mathrm{pb}}) = \omega_{\mathrm{pb}}^2\omega_{\mathrm{po}}^2. \tag{5.110}$$

This is an alternative way of displaying the approximate roots of eqn (5.106). In general, the beam population is small, so that

$$\xi \ll 1. \tag{5.111}$$

Then the right-hand side of eqn (5.110) is small compared to ω_{po}^4. This restricts the magnitude of the left-hand side, so that at least one among the four factors must also be small. Either

$$\omega \simeq \pm\omega_{\mathrm{po}}, \tag{5.112}$$

corresponding to eqn (5.107); or

$$\omega \simeq kv_0 \pm \omega_{\mathrm{pb}}, \tag{5.113}$$

corresponding to eqn (5.108).

Now let us examine what happens when the two families of waves overlap in frequency. The condition for frequency overlap can be expressed in terms of the critical wavenumber k_{c}, defined by

$$k_{\mathrm{c}} = \omega_{\mathrm{po}}/v_0. \tag{5.114}$$

When $k = k_c$, eqn (5.113) gives $\omega \simeq \omega_{po} \pm \omega_{pb}$ for beam-supported waves. By eqns (5.105) and (5.111), this frequency is close to ω_{po}, which by eqn (5.112) is the characteristic frequency of background-supported waves. Thus,

$$k \simeq k_c \tag{5.115}$$

is the condition for frequency resonance to be possible between waves from the two families. We now seek solutions of eqn (5.109) which have the form

$$\omega = \omega_{po} + \eta, \qquad |\eta| \ll \omega_{po}. \tag{5.116}$$

For the case $k = k_c$, substitution of eqns (5.116) and (5.114) into eqn (5.109) yields

$$\eta^3 - \omega_{pb}^2 \eta = \omega_{pb}^2 \omega_{po}/2. \tag{5.117}$$

If we assume

$$|\eta| \gg \omega_{pb}, \tag{5.118}$$

which we shall check subsequently for consistency, we may neglect $\omega_{pb}^2 \eta$ compared to η^3 in eqn (5.117), leaving

$$\eta^3 = \omega_{pb}^2 \omega_{po}/2 = \xi \omega_{pe}^3/2. \tag{5.119}$$

In the final expression in eqn (5.119), we have used eqns (5.104), (5.105), and (5.111). Now unity has three complex cube roots $\{1, \exp(2i\pi/3), \exp(4i\pi/3)\}$, and eqn (5.119) has three corresponding roots. These are

$$\eta_0 = \left(\frac{\xi}{2}\right)^{\frac{1}{3}} \omega_{pe}, \tag{5.120}$$

$$\eta_1 = \eta_0 \left(-\frac{1}{2} + i\frac{\sqrt{3}}{2}\right), \tag{5.121}$$

$$\eta_2 = \eta_0 \left(-\frac{1}{2} - i\frac{\sqrt{3}}{2}\right). \tag{5.122}$$

Combining eqns (5.120) to (5.122) with (5.116), the first root has a purely real frequency. The effect of the linear frequency resonance for this root is to introduce a small frequency shift η_0 with respect to the background plasma frequency ω_{po}. The second and third roots include an imaginary term, which we write as $\pm i\gamma$ where

$$\gamma = \left(\frac{\sqrt{3}}{2}\right)\eta_0 = \frac{\sqrt{3}}{2^{\frac{4}{3}}} \xi^{\frac{1}{3}} \omega_{pe}. \tag{5.123}$$

Following the discussion of eqns (5.60) and (5.61), the root $\omega = \omega_{po} + \eta_2$ of eqn (5.109) grows exponentially at the rate γ. Note that, at resonance, the waves supported by the background and beam plasmas are indistinguishable, because their wavenumber k_c and frequency $\omega_{po} + \eta_2$ are identical. We have shown that this collective beam-plasma electrostatic oscillation can grow exponentially. Before discussing this instability in greater detail, let us briefly confirm the consistency of our results, eqns (5.120) to (5.122). First, using eqn (5.111), it is clear that $|\eta| \ll \omega_{\text{po}}$, as required at eqn (5.116). Second, using eqns (5.111) and (5.105), we are indeed consistent with eqn (5.118).

The phenomenon that we have identified is known as the two-stream instability, the two streams being the beam ($v_z = v_0$) and the background plasma ($v_z = 0$). It is a mechanism by which the beam and background components of the plasma may interact, mediated by resonance between the collective oscillations that they support. The instability rests on linear frequency resonance, which occurs when $k = k_{\text{c}}$ defined by eqn (5.114). It results in the growth of the average electrostatic field energy that is associated with the resonant oscillation. There is only one possible source for this energy: the free kinetic energy associated with the directed motion of the beam electrons. In order to examine this flow of energy, we must discuss these wave phenomena at the level of particle dynamics.

Let us denote the background electron population, whose velocity is zero, by subscript a, and the beam electron population, whose equilibrium velocity is v_0 in the z-direction, by subscript b. As before, we are concerned only with density and velocity perturbations in the z-direction. Under these perturbations, the number of electrons in the two populations is separately conserved. Applying eqn (5.4), we have

$$\frac{\partial}{\partial t}\,\delta n_{\text{a}} + \frac{\partial}{\partial z}\,\{(n_{\text{a}} + \delta n_{\text{a}})\,\delta v_{\text{a}}\} = 0, \tag{5.124}$$

$$\frac{\partial}{\partial t}\,\delta n_{\text{b}} + \frac{\partial}{\partial z}\,\{(n_{\text{b}} + \delta n_{\text{b}})(v_0 + \delta v_{\text{b}})\} = 0. \tag{5.125}$$

We assume that the perturbed quantities vary as $\exp(\mathrm{i}kz - \mathrm{i}\omega t)$, and neglect all quadratic products of perturbed quantities. This linearization of eqns (5.124) and (5.125) yields

$$\omega \delta n_{\text{a}} = k n_{\text{a}}\,\delta v_{\text{a}}, \tag{5.126}$$

$$(\omega - k v_0)\,\delta n_{\text{b}} = k n_{\text{b}}\,\delta v_{\text{b}}. \tag{5.127}$$

The individual electrons in the two populations obey the force equations

$$\frac{\partial}{\partial t}\,\delta v_{\text{a}} = -\frac{e}{m}\,E, \tag{5.128}$$

$$\frac{\partial}{\partial t}\,\delta v_{\text{b}} + v_0 \frac{\partial}{\partial z}\,\delta v_{\text{b}} = -\frac{e}{m}\,E. \tag{5.129}$$

It follows that

$$\omega\,\delta v_{\mathrm{a}} = -\frac{\mathrm{i}e}{m}E, \tag{5.130}$$

$$(\omega - kv_0)\,\delta v_{\mathrm{b}} = -\frac{\mathrm{i}e}{m}E. \tag{5.131}$$

The electric field E arises from the combined charge density perturbations of the two electron populations. Using Poisson's equation (I.1), we have

$$\frac{\partial E}{\partial z} = -\frac{e}{\varepsilon_0}(\delta n_{\mathrm{a}} + \delta n_{\mathrm{b}}), \tag{5.132}$$

which yields

$$kE = \frac{\mathrm{i}e}{\varepsilon_0}(\delta n_{\mathrm{a}} + \delta n_{\mathrm{b}}). \tag{5.133}$$

This closes the set of five equations, eqns (5.126), (5.127), (5.130), (5.131), and (5.133) for the five perturbed quantities δn_{a}, δn_{b}, δv_{a}, δv_{b}, and E, that describe the beam-plasma system.

We first eliminate E from eqns (5.130) and (5.131) using eqn (5.133):

$$\omega\,\delta v_{\mathrm{a}} = \frac{e^2}{m\varepsilon_0 k}(\delta n_{\mathrm{a}} + \delta n_{\mathrm{b}}), \tag{5.134}$$

$$(\omega - kv_0)\,\delta v_{\mathrm{b}} = \frac{e^2}{m\varepsilon_0 k}(\delta n_{\mathrm{a}} + \delta n_{\mathrm{b}}). \tag{5.135}$$

Substituting eqn (5.134) into eqn (5.126), and eqn (5.135) into eqn (5.127), we obtain

$$\omega\,\delta n_{\mathrm{a}} = \frac{n_{\mathrm{a}}e^2}{m\varepsilon_0\omega}(\delta n_{\mathrm{a}} + \delta n_{\mathrm{b}}), \tag{5.136}$$

$$(\omega - kv_0)\,\delta n_{\mathrm{b}} = \frac{n_{\mathrm{b}}e^2}{m\varepsilon_0(\omega - kv_0)}(\delta n_{\mathrm{a}} + \delta n_{\mathrm{b}}). \tag{5.137}$$

By analogy with eqns (5.104) and (5.105), it is convenient to define separate plasma frequencies for the two electron populations:

$$\omega_{\mathrm{pa,b}}^2 = \frac{n_{\mathrm{a,b}}e^2}{m\varepsilon_0}. \tag{5.138}$$

Then eqns (5.136) and (5.137) can be written

$$(\omega^2 - \omega_{\mathrm{pa}}^2)\,\delta n_{\mathrm{a}} = \omega_{\mathrm{pa}}^2\,\delta n_{\mathrm{b}}, \tag{5.139}$$

$$\{(\omega - kv_0)^2 - \omega_{\mathrm{pb}}^2\}\,\delta n_{\mathrm{b}} = \omega_{\mathrm{pb}}^2\,\delta n_{\mathrm{a}}. \tag{5.140}$$

It follows that the dispersion relation is

$$(\omega^2 - \omega_{\mathrm{pa}}^2)\{(\omega - kv_0)^2 - \omega_{\mathrm{pb}}^2\} = \omega_{\mathrm{pa}}^2\omega_{\mathrm{pb}}^2. \tag{5.141}$$

This is identical to eqn (5.109), which was obtained using the dielectric approach. We are now in a position to study the dynamical properties of the solutions of this dispersion relation.

Let us examine the two wave branches that are supported by the beam. Their approximate dispersion relation has already been given at eqn (5.113). First, corresponding to the plus sign in eqn (5.113), there is the fast wave, so-called because its phase velocity is slightly greater than the velocity of the beam:

$$\frac{\omega_{\mathrm{f}}}{k} = v_0 + \frac{\omega_{\mathrm{pb}}}{k}. \tag{5.142}$$

Second, corresponding to the minus sign in eqn (5.113), there is the slow wave, whose phase velocity is slightly less than that of the beam:

$$\frac{\omega_{\mathrm{s}}}{k} = v_0 - \frac{\omega_{\mathrm{pb}}}{k}. \tag{5.143}$$

Thus, the phase velocities of the two modes supported by the beam differ by $2\omega_{\mathrm{pb}}/k$, but have the same sign, corresponding to phase motion in the same direction. The difference in phase velocity is determined by the phase relation of the density and velocity perturbations of the beam, as follows. Substituting eqn (5.142) into eqn (5.127), we see that for the fast wave,

$$\frac{\omega_{\mathrm{pb}}}{k}\,\delta n_{\mathrm{b}} = n_{\mathrm{b}}\,\delta v_{\mathrm{b}}; \tag{5.144}$$

in this case, δn_{b} and δv_{b} are in phase. For the slow wave, substituting eqn (5.143) into (5.127) gives

$$-\frac{\omega_{\mathrm{pb}}}{k}\,\delta n_{\mathrm{b}} = n_{\mathrm{b}}\,\delta v_{\mathrm{b}}, \tag{5.145}$$

so that δn_{b} and δv_{b} are out of phase by π. This degree of freedom in the phase of density and velocity perturbations exists only for the beam, and not for the background population. We note that this is the second time that we have seen fast and slow branches of a wave arising from the phase relation of the two perturbations that comprise the wave; the example of fast and slow magnetosonic waves was discussed after eqn (4.86).

The existence of the waves alters the total momentum density of the beam. This quantity can be written

$$p_{\mathrm{b}} = m(n_{\mathrm{b}} + \delta n_{\mathrm{b}})(v_0 + \delta v_{\mathrm{b}}). \tag{5.146}$$

Now δn_b and δv_b are oscillatory quantities associated with the waves. We wish to calculate the average of p_b over many oscillations. Denoting the averaging process by $\langle\ \rangle$, it is clear that, since $\langle \sin \omega t \rangle = 0$ and $\langle \sin^2 \omega t \rangle = \frac{1}{2}$,

$$\langle \delta n_b \rangle = \langle \delta v_b \rangle = 0, \tag{5.147}$$

whereas $\langle \delta n_b^2 \rangle$ and $\langle \delta v_b^2 \rangle$ are positive. Returning to eqn (5.146), it follows that the average beam momentum is

$$\langle p_b \rangle = m(n_b v_0 + \langle \delta n_b\, \delta v_b \rangle). \tag{5.148}$$

The equilibrium value of the beam momentum density is

$$p_{bo} = m n_b v_0. \tag{5.149}$$

It follows from eqns (5.148) and (5.149) that the change in average beam momentum density arising from the presence of the wave is

$$\langle p_b \rangle - p_{bo} = m \langle \delta n_b\, \delta v_b \rangle. \tag{5.150}$$

If the fast wave is present, eqns (5.144) and (5.150) yield

$$\langle p_b \rangle - p_{bo} = \frac{m \omega_{pb}}{n_b k} \langle \delta n_b^2 \rangle, \tag{5.151}$$

which is a positive quantity. Thus, the beam is on average in a state of increased momentum when the fast wave is present. The excitation of the fast wave accordingly requires an energy input. If the slow wave is present, eqns (5.145) and (5.150) yield

$$\langle p_b \rangle - p_{bo} = -\frac{m \omega_{pb}}{k n_b} \langle \delta n_b^2 \rangle. \tag{5.152}$$

This is negative. The beam is in a state of reduced momentum when the slow wave is present, and the average value of this reduction is related to the amplitude of the slow wave by eqn (5.152). Conversely, the excitation of the slow wave is a means of reducing the directed kinetic energy of the beam. For this reason, the slow wave is classified as a negative energy wave. The physical origin for this effect lies in the phase relation derived at eqn (5.145). At each peak of velocity associated with the slow wave, the density is reduced, and vice versa: see Fig. 5.12. Thus, relative to unperturbed levels, there are more electrons with reduced velocity. In studying Fig. 5.12, it should be noted that we cannot reduce v_0 to zero, while leaving the rest of the diagram unchanged. This is because the phase difference of π between δn and δv is possible only if a distinct beam exists, so that v_0 cannot be zero.

Although a beam with a slow wave present is in a lower energy state than when it is unperturbed, the beam cannot spontaneously excite the slow wave and enter the lower energy state. There must exist a

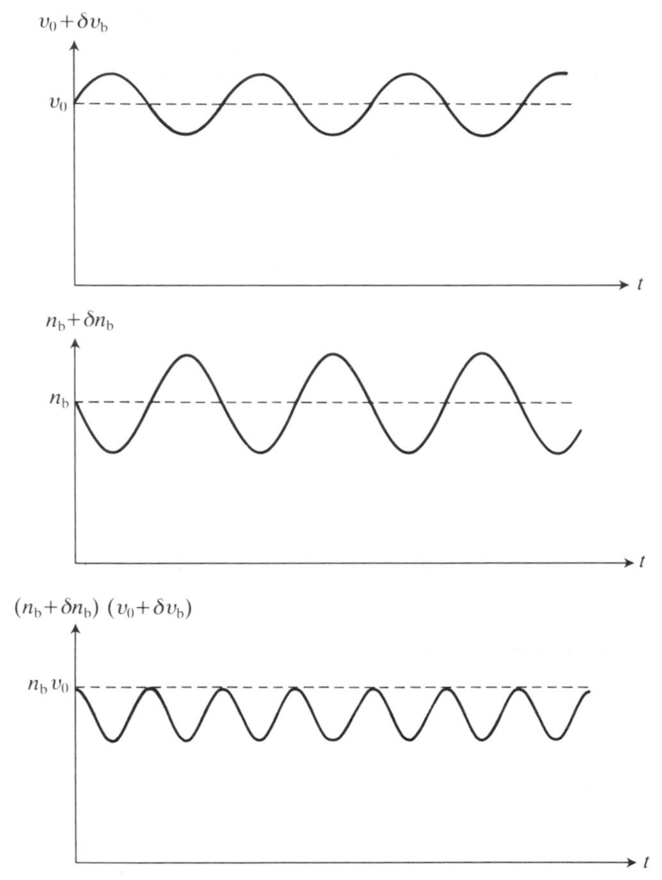

Fig. 5.12 Velocity–density relation for the slow wave.

mechanism that will transfer the energy released from the beam to the rest of the plasma. The required coupling between the beam and the background plasma can occur when there is frequency resonance between the collective oscillations supported by the two populations. Consider first the case of exact resonance between the background wave $\omega = \omega_{pa}$ and the slow wave $\omega = kv_0 - \omega_{pb}$. This requires the wavenumber to have the critical value

$$k_{c1} = (\omega_{pa} + \omega_{pb})/v_0. \tag{5.153}$$

Note that k_{c1} differs slightly from k_c defined by eqn (5.114). Following our earlier approach, we seek solutions to eqn (5.141) that have the form

$$\omega = \omega_{pa} + \eta, \qquad |\eta| \ll \omega_{pa}, \tag{5.154}$$

when $k = k_{c1}$. We obtain

$$\eta^3 - 2\omega_{pb}\eta^2 = \omega_{pa}\omega_{pb}^2/2, \tag{5.155}$$

which differs slightly from eqn (5.117). If we again adopt the approximation eqn (5.118), we recover the results given in eqns (5.120) to (5.123). We have thus met our objective of outlining the energy flow that occurs in the two-stream instability. The growing resonant oscillation of the electrostatic field, which is supported by both the beam and background populations, includes a slow wave component whose excitation reduces the kinetic energy of the beam.

The higher-order differences between eqns (5.117) and (5.155) arise from the fact that at $k = k_{c1}$, there is exact resonance between the background and slow waves, whereas at $k = k_c$, the background wave has the same degree of approximate resonance with both the fast and slow waves. That is, using eqns (5.143) and (5.153),

$$\omega_s = \omega_{pa} \tag{5.156}$$

when $k = k_{c1}$; whereas when $k = k_c$,

$$\omega_f = \omega_{pa} + \omega_{pb} \tag{5.157}$$

$$\omega_s = \omega_{pa} - \omega_{pb} \tag{5.158}$$

using eqns (5.117), (5.142), and (5.143), and writing ω_{pa} for ω_{po} since we have established that they are equivalent. When the higher-order terms are included, it can be shown that eqn (5.117) yields a lower growth rate than eqn (5.155). Physically, this occurs because the growth of the approximately resonant, positive energy fast wave acts to reduce the rate of energy loss from the beam.

Finally, let us consider a mechanism suggested by Lashmore-Davies, for the termination of the instability of the slow wave with $k = k_{c1}$. We know from eqn (5.151) that the mean velocity of the beam is reduced as the amplitude of the slow wave grows. Using eqn (5.152), we may write the change in beam velocity as

$$\delta v_0 = \frac{\langle p_b \rangle - p_{bo}}{m n_b} = -\frac{\omega_{pb}}{k} \frac{\langle \delta n_b^2 \rangle}{n_b^2}. \tag{5.159}$$

It is the beam velocity, which is now to be written $v_0 + \delta v_0$, that determines the frequency of the slow wave through eqns (5.140) and (5.143). The frequency of a slow wave with given wavenumber therefore changes as the instability proceeds. Using eqns (5.143) and (5.159), we have

$$\begin{aligned} \omega_s &= k(v_0 + \delta v_0) - \omega_{pb} \\ &= k v_0 - \omega_{pb}\left(1 + \frac{\langle \delta n_b^2 \rangle}{n_0}\right). \end{aligned} \tag{5.160}$$

Note that this frequency shift is a non-linear effect. It depends on the square of the amplitude of the wave. We consider the slow wave with wavenumber k_{c1} defined by eqn (5.153), which initially satisfies the exact resonance condition eqn (5.156). Then eqn (5.160) gives

$$\omega_s = \omega_{pa} - \frac{\langle \delta n_b^2 \rangle}{n_b^2} \omega_{pb}. \tag{5.161}$$

Initially, δn_b is negligible, and eqn (5.161) reduces to eqn (5.156). As $\langle \delta n_b^2 \rangle$ grows due to instability, eqn (5.161) shows the corresponding increase in frequency mismatch between the slow wave with wavevector k_{c1} and the background plasma wave. Eventually, the development of the instability will destroy the resonance condition, terminating the growth of the wave. As this occurs, another slow wave with a different wavenumber will enter resonance and become unstable.

In this section, we have seen how consideration of non-linear quantities, such as $\langle \delta n_b \, \delta v_b \rangle$ and $\langle \delta n_b^2 \rangle$, leads to a better understanding of the essentially linear phenomenon of two-stream instability. We shall go further in this direction in the next chapter, and construct a more general approach to non-linear plasma dynamics.

Exercises

5.1. Fluid quantities of the type discussed in Chapter 4 can be identified with moments of the distribution function. For example, the electron fluid has mass density $\rho(\boldsymbol{x}, t) = \int mf \, d^3\boldsymbol{v}$, and its bulk fluid velocity is equal to the mean electron velocity $(1/\rho) \int m\boldsymbol{v}f \, d^3\boldsymbol{v}$, where the integrations are over all velocities. By considering the integral of the Vlasov equation, eqn (5.9), over all velocities, show that the electron fluid satisfies a continuity equation of the form eqn (4.1).

5.2. The equilibrium velocity distribution of some types of plasma particle is not necessarily a Maxwellian. For example, if a distinct population of energetic ions is created in a tokamak plasma, it can often be described by the 'slowing-down distribution':

$$f_0(v) = \begin{cases} \dfrac{A}{v_c^3 + v^3} & v \leqslant v_b \\ 0 & v > v_b \, . \end{cases}$$

Here, v_b is the velocity at which the energetic ions are born, for example as a result of fusion reactions; v_c is a constant which is determined by the collisionality of the plasma; and A is a further constant: obtain a formula for it.

5.3. The kinetic description of galaxies has many similarities with that of plasmas. Because collisions between stars in galaxies are very rare, the evolution of the distribution of stars in phase space can be described by a continuity equation which has the form of eqn (5.5). Each star interacts with the rest of the galaxy through the local gravitational potential $\Phi(\boldsymbol{x})$.

(a) Obtain the Vlasov equation for the stars in a galaxy, which is analogous to eqn (5.9) for the particles in a plasma.

(b) Defining the number density of stars in space $\nu = \int f \, d^3\boldsymbol{v}$, and the mean stellar velocity $\bar{\boldsymbol{v}}(\boldsymbol{x}) = (1/\nu) \int \boldsymbol{v} f \, d^3\boldsymbol{v}$, obtain the continuity equation for the galaxy.

(c) Using index notation, of the type employed in Section 4.2, show that

$$\frac{\partial}{\partial t}(\nu \bar{v}_j) + \frac{\partial}{\partial x_i}(\nu \overline{v_i v_j}) + \nu \frac{\partial \Phi}{\partial x_j} = 0,$$

where $\nu \overline{v_i v_j} = \int v_i v_j f \, d^3\boldsymbol{v}$.

Solutions are on pages 153 *to* 155.

6

Non-linear plasma physics

6.1 Two-fluid theory and the Zakharov equations

We shall now consider the connection between the two major ways of describing plasma that have been used so far. First, we considered the dynamics of a single particle in the plasma, which leads directly to a simple model for the dielectric properties of the plasma. By including the single-particle dynamics of ions as well as of electrons, it is possible to predict the normal modes of the plasma over a wide range of frequencies. There is also the magnetohydrodynamic approach. By treating the plasma as a magnetized conducting fluid, whose inertia is provided by the mass of the ions, we can describe the bulk stability of the plasma and its lower-frequency normal modes—those modes which do not involve the movement of electrons independently of ions. The question now arises: what is the connection between the two ways of describing the plasma? For example, how is a Langmuir wave affected by the additional presence of a much lower-frequency magnetohydrodynamic wave? In this case, any changes in the local ion number density associated with the magnetohydrodynamic wave will alter the restoring Coulomb force that an electron experiences when it is displaced. By the discussion in Chapter 1, this will alter the local electron plasma frequency, which will affect the high-frequency Langmuir wave. Previously, we have treated the Langmuir wave and the magnetohydrodynamic wave as two distinct linear normal modes of the plasma. We have now given a simple physical description of the way in which these two linear normal modes may interact. This is an essentially non-linear problem.

It is possible to address the questions that we have raised by using the two-fluid, two-timescale theory that we shall discuss in this section. The plasma is modelled in terms of two intermingled fluids, the electron fluid and the ion fluid. Two timescales, fast and slow, arise from the wide difference between the mass of an electron and that of an ion. In response to a given force, the acceleration of an electron will be much greater than that of an ion, because its mass is roughly two thousand times less. For this reason, the characteristic frequencies of electron motion—for example the electron plasma frequency and cyclotron frequency—are much higher than the corresponding frequencies for ions. The reciprocals of the characteristic electron frequency and of the characteristic ion frequency determine the fast and slow timescales respectively.

As an example of two-fluid, two-timescale theory, we shall consider the behaviour of electrons and ions in a plasma with no magnetic field present. The absence of a magnetic field will keep the equations relatively simple without restricting us conceptually. The linear normal modes of this system are known to be fast-timescale Langmuir waves and slow-timescale ion acoustic waves. We shall identify these modes using two-fluid theory, and see how they interact. First, the number density of the ions will be written

$$n_{\mathrm{i}} = n_0 + \delta n_{\mathrm{i}}. \tag{6.1}$$

Here n_0 denotes the time-independent equilibrium value, and δn_{i} is a slow-timescale perturbation. The electron number density is

$$n_{\mathrm{e}} = n_0 + \delta n_{\mathrm{e}} + \tilde{n}_{\mathrm{e}}. \tag{6.2}$$

Here $\tilde{n}_{\mathrm{e}}$ is the fast-timescale perturbation, which oscillates at some frequency close to the electron plasma frequency ω_{pe}. Its average value over many such oscillations, denoted by $\langle \tilde{n}_{\mathrm{e}} \rangle$, is zero. It follows that the value of the electron number density, averaged over many fast oscillations, is

$$\langle n_{\mathrm{e}} \rangle = n_0 + \delta n_{\mathrm{e}}. \tag{6.3}$$

This quantity is still subject to the slow-timescale perturbation δn_{e}. The quantities δn_{i} and δn_{e} appearing in eqns (6.1) and (6.3) are nearly identical. This is because the mobility of the electrons is so high that their neutralizing response to electric fields is sufficiently rapid to keep the plasma almost exactly electrically neutral throughout any slow-timescale perturbation. We express this in the quasineutrality condition

$$\delta n_i \simeq \delta n_{\mathrm{e}}, \tag{6.4}$$

and we shall write δn for both δn_{i} and δn_{e} except in circumstances where it is important to include the slight deviation from charge neutrality on the slow timescale. The equilibrium number density n_0 of the electrons is exactly equal to that of the ions. Deviations from charge neutrality are described by $n_{\mathrm{e}} - n_{\mathrm{i}}$, so that Poisson's equation eqn (I.1) becomes

$$\nabla \cdot \boldsymbol{E} = -e\tilde{n}_{\mathrm{e}}/\varepsilon_0 - e(\delta n_{\mathrm{e}} - \delta n_{\mathrm{i}})/\varepsilon_0. \tag{6.5}$$

This illustrates the two distinct sources of the electric field $\boldsymbol{E}$. To reflect this, we shall divide $\boldsymbol{E}$ into two parts as follows:

$$\boldsymbol{E} = \tilde{\boldsymbol{E}} + \boldsymbol{E}_{\mathrm{slow}}, \tag{6.6}$$

$$\nabla \cdot \tilde{\boldsymbol{E}} = -e\tilde{n}_{\mathrm{e}}/\varepsilon_0, \tag{6.7}$$

$$\nabla \cdot \boldsymbol{E}_{\mathrm{slow}} = -e(\delta n_{\mathrm{e}} - \delta n_{\mathrm{i}})/\varepsilon_0. \tag{6.8}$$

The rapid oscillations in electron number density $\tilde{n}_{\mathrm{e}}$ are the source of the fast-timescale component $\tilde{\boldsymbol{E}}$. Averaged over many such oscillations, the electric field has the value $\langle \boldsymbol{E} \rangle = \boldsymbol{E}_{\text{slow}}$, whose source is the slow-timescale non-neutrality that we have discussed.

We shall write the velocity of the electron fluid as

$$\boldsymbol{v}_{\text{e}} = \delta \boldsymbol{v}_{\text{e}} + \tilde{\boldsymbol{v}}_{\text{e}}, \tag{6.9}$$

and that of the ion fluid as

$$\boldsymbol{v}_{\text{i}} = \delta \boldsymbol{v}_{\text{i}}. \tag{6.10}$$

Here $\tilde{\boldsymbol{v}}_{\text{e}}$ is the fast component of electron motion, to which there is no ion counterpart. On the slower timescale, averaged over many fast oscillations, the electron and ion fluids move almost together with $\delta \boldsymbol{v}_{\text{e}} \simeq \delta \boldsymbol{v}_{\text{i}}$. At equilibrium, both fluids are at rest, so that the equilibrium velocity is zero. The total plasma current is

$$\boldsymbol{J} = -e(n_{\text{e}} \boldsymbol{v}_{\text{e}} - n_{\text{i}} \boldsymbol{v}_{\text{i}}). \tag{6.11}$$

Combining eqns (6.1), (6.2), (6.4), (6.9) and (6.10), this gives to leading order

$$\boldsymbol{J} = -e\{(n_0 + \delta n)\tilde{\boldsymbol{v}}_{\text{e}} + \tilde{n}_{\text{e}} \tilde{\boldsymbol{v}}_{\text{e}}\}. \tag{6.12}$$

We shall shortly discuss how to deal with the product of two fast timescale quantities, such as $\tilde{n}_{\text{e}} \tilde{\boldsymbol{v}}_{\text{e}}$ in eqn (6.12). For immediate purposes, however, we shall neglect $\tilde{n}_{\text{e}} \tilde{\boldsymbol{v}}_{\text{e}}$ in eqn (6.12) on the grounds that $\delta n \gg \tilde{n}_{\text{e}}$. This assumption follows from the fact that the amplitude of $\tilde{n}_{\text{e}}$ is limited by the strong Coulomb force, which does not affect the charge-neutral perturbation δn. Then the fast-timescale plasma current is, to leading order,

$$\tilde{\boldsymbol{J}} = -e(n_0 + \delta n)\tilde{\boldsymbol{v}}_{\text{e}}. \tag{6.13}$$

Now recall the magnetic source equation eqn (I.9), which gives $\nabla \times \tilde{\boldsymbol{H}} = \tilde{\boldsymbol{J}} + \varepsilon_0(\partial \tilde{\boldsymbol{E}} / \partial t)$ for the fast-timescale fields and current. Then by eqn (6.13),

$$\nabla \times \tilde{\boldsymbol{B}} = -\mu_0 e(n_0 + \delta n)\tilde{\boldsymbol{v}}_{\text{e}} + \frac{1}{c^2} \frac{\partial \tilde{\boldsymbol{E}}}{\partial t}. \tag{6.14}$$

Here, we have used the fact that $\boldsymbol{B} = \mu_0 \boldsymbol{H}$ and $\mu_0 \varepsilon_0 = 1/c^2$. Also, we have $\nabla \times \tilde{\boldsymbol{E}} = -\partial \tilde{\boldsymbol{B}} / \partial t$ by eqn (I.4). Operating on eqn (6.14) with $-\partial/\partial t$, we obtain

$$\nabla \times (\nabla \times \tilde{\boldsymbol{E}}) + \frac{1}{c^2} \frac{\partial^2 \tilde{\boldsymbol{E}}}{\partial t^2} = \mu_0 e \frac{\partial}{\partial t} \{(n_0 + \delta n)\tilde{\boldsymbol{v}}_{\text{e}}\}. \tag{6.15}$$

We can simplify the right-hand side of eqn (6.15), if we note that

$$\frac{\partial}{\partial t} \{(n_0 + \delta n)\tilde{\boldsymbol{v}}_{\text{e}}\} = \tilde{\boldsymbol{v}}_{\text{e}} \frac{\partial}{\partial t} \delta n + (n_0 + \delta n) \frac{\partial}{\partial t} \tilde{\boldsymbol{v}}_{\text{e}}. \tag{6.16}$$

Now the time derivative of a given quantity is by definition smaller if the quantity is slowly varying than if it is rapidly varying. We accordingly neglect the first term on the right-hand side of eqn (6.16). Substituting the remaining term into eqn (6.15), and multiplying through by $c^2 = 1/\mu_0\varepsilon_0$, we obtain

$$c^2\nabla\times(\nabla\times\tilde{\boldsymbol{E}})+\frac{\partial^2\tilde{\boldsymbol{E}}}{\partial t^2}=\frac{e}{\varepsilon_0}(n_0+\delta n)\frac{\partial\tilde{\boldsymbol{v}}_e}{\partial t}. \tag{6.17}$$

Equations (6.7) and (6.17) represent a restatement of Maxwell's equations in terms of the fast timescale electric field amplitude $\tilde{\boldsymbol{E}}$ and our two-fluid, two-timescale variables δn, $\tilde{n}_e$ and $\tilde{\boldsymbol{v}}_e$. In order to progress further, we require an additional input from fluid theory. This is provided by the equation of motion of the electron fluid:

$$m\frac{\partial\boldsymbol{v}_e}{\partial t}+m(\boldsymbol{v}_e\cdot\nabla)\boldsymbol{v}_e=-e\boldsymbol{E}-\gamma_e\frac{k_B T_e}{n_e}\nabla n_e. \tag{6.18}$$

The final term on the right-hand side is the pressure term that is obtained from simple kinetic theory. We substitute for n_e and $\boldsymbol{v}_e$ in eqn (6.18) using eqns (6.2) and (6.9). To leading order the fast-timescale equation of motion for the electron fluid is

$$\frac{\partial\tilde{\boldsymbol{v}}_e}{\partial t}=-\frac{e}{m}\tilde{\boldsymbol{E}}-\frac{3v_{Te}^2}{n_0}\nabla\tilde{n}_e. \tag{6.19}$$

Here, $v_{Te}=(k_B T_e/m)^{\frac{1}{2}}$ is the electron thermal velocity. We have set $\gamma_e=3$ for the fast-timescale perturbations of the electron fluid. This follows from the assumption that these perturbations are adiabatic: that is, the oscillations are so rapid that there is not time for the electrons to move and transport energy down the temperature gradients associated with the wave. Later, when we consider slow-timescale waves involving both electrons and ions, we shall set $\gamma_e=1$ and $\gamma_i=3$. This corresponds to isothermal perturbations of the electrons, which have time to move rapidly to equalize any electron temperature gradients, thus preventing their occurrence in the wave; meanwhile the less mobile ions undergo adiabatic perturbation. Equation (6.19) provides a further link between the fast-timescale electron fluid variables $\tilde{n}_e$ and $\tilde{\boldsymbol{v}}_e$ and the fast-timescale electric field $\tilde{\boldsymbol{E}}$. Our aim now is to use $\tilde{\boldsymbol{E}}$ as the only fast-timescale variable. We first eliminate $\tilde{n}_e$ by substituting from eqn (6.7) in eqn (6.19):

$$\frac{\partial\tilde{\boldsymbol{v}}_e}{\partial t}=-\frac{e}{m}\tilde{\boldsymbol{E}}+\frac{3\varepsilon_0 v_{Te}^2}{n_0 e}\nabla(\nabla\cdot\tilde{\boldsymbol{E}}). \tag{6.20}$$

Then we combine eqns (6.17) and (6.20), and use eqn (1.6), to obtain

$$\frac{\partial^2\tilde{\boldsymbol{E}}}{\partial t^2}+\omega_{pe}^2\left(1+\frac{\delta n}{n_0}\right)\tilde{\boldsymbol{E}}-3v_{Te}^2\nabla(\nabla\cdot\tilde{\boldsymbol{E}})+c^2\nabla\times(\nabla\times\tilde{\boldsymbol{E}})=0. \tag{6.21}$$

This equation has many attractive features. First, it is a non-linear equation; the non-linearity occurs in the product of the slow-timescale density perturbation δn with the rapidly oscillating electric field $\tilde{\boldsymbol{E}}$. As we expected from our introductory discussion, δn arises in this equation because it changes the local electron plasma frequency from its equilibrium value ω_{pe} to $\omega_{pe}(1+\delta n/n_0)^{\frac{1}{2}}$. This term originates in the correction to the fast-timescale plasma current that is caused by the slow-timescale density perturbation, which is contained in eqn (6.13). We have neglected the other non-linear term generated by eqns (6.17) and (6.20), which is proportional to $\delta n \boldsymbol{\nabla}(\boldsymbol{\nabla}\cdot\tilde{\boldsymbol{E}})$, on grounds of smallness. The quantity $3v_{Te}^2\boldsymbol{\nabla}(\boldsymbol{\nabla}\cdot\tilde{\boldsymbol{E}})$ is a small thermal correction to the cold plasma term $\omega_{pe}^2\tilde{\boldsymbol{E}}$. Suppose now that we linearize eqn (6.21), that is, we neglect the non-linear term $\omega_{pe}^2\delta n\tilde{\boldsymbol{E}}/n_0$. In this limit, eqn (6.21) will give us the linear dispersion relation for electrostatic waves. Let us take the space and time dependence of $\tilde{\boldsymbol{E}}$ to be proportional to $\exp(\mathrm{i}\boldsymbol{k}\cdot\boldsymbol{r}-\mathrm{i}\omega t)$, with $\boldsymbol{k}$ parallel to $\tilde{\boldsymbol{E}}$. Substituting this in eqn (6.21) gives

$$\omega^2=\omega_{pe}^2+3v_{Te}^2k^2=\omega_{pe}^2(1+3k^2/k_D^2), \tag{6.22}$$

where as usual k_D is the Debye wavenumber, $k_D=\omega_{pe}/v_{Te}$. Equation (6.22) is identical to eqns (1.21) and (5.38). This is a useful check for the consistency of our fluid description. Tracing the origin of the term $3k^2/k_D^2$ back through eqns (6.20), (6.19), and (6.18), we see that it comes from the pressure term in the equation of motion for the electron fluid. This is what we would expect on the basis of our discussion in Section 1.3.

Now let us return to the development of the nonlinear theory. We have not yet considered the slow-timescale motion of the two fluids. For example, the velocity of the ion fluid has entered only in eqn (6.11), where its contribution is immediately cancelled by slow-timescale electron terms. Also, we have not yet discussed in general how to deal with terms that are quadratic in fast-timescale quantities, such as $\tilde{n}_e\tilde{\boldsymbol{v}}_e$ in eqn (6.12). Our first step will be to introduce a convenient formalism. In particular, let us consider the fast-timescale electric field $\tilde{\boldsymbol{E}}$. Its time evolution has two widely separated components: oscillation at or near the electron plasma frequency ω_{pe}; and a much slower variation of the amplitude of oscillation. We therefore choose to write

$$\tilde{\boldsymbol{E}}(\boldsymbol{r},t)=\bar{\boldsymbol{E}}_c(\boldsymbol{r},t)\cos\omega_{pe}t+\bar{\boldsymbol{E}}_s(\boldsymbol{r},t)\sin\omega_{pe}t. \tag{6.23}$$

This equation defines $\bar{\boldsymbol{E}}_c(\boldsymbol{r},t)$ and $\bar{\boldsymbol{E}}_s(\boldsymbol{r},t)$ as the slowly varying and independent amplitudes of the components of the electric field that oscillate as $\cos\omega_{pe}t$ and $\sin\omega_{pe}t$ respectively. Conversely, insofar as $\tilde{\boldsymbol{E}}(\boldsymbol{r},t)$ does not oscillate precisely as $\cos\omega_{pe}t$ and $\sin\omega_{pe}t$, the additional time-dependence is contained in $\bar{\boldsymbol{E}}_c(\boldsymbol{r},t)$ and $\bar{\boldsymbol{E}}_s(\boldsymbol{r},t)$. Every quantity that we have used so far is a real—as opposed to complex—quantity. It is now convenient to define a complex variable

$$\bar{\boldsymbol{E}}(\boldsymbol{r},t)=\bar{\boldsymbol{E}}_c(\boldsymbol{r},t)+\mathrm{i}\bar{\boldsymbol{E}}_s(\boldsymbol{r},t). \tag{6.24}$$

Then we can write the fast-timescale electric field $\tilde{\boldsymbol{E}}(\boldsymbol{r}, t)$ in the following form, which is equivalent to eqn (6.23):

$$\begin{aligned}\tilde{\boldsymbol{E}}(\boldsymbol{r}, t) &= \tfrac{1}{2}\{\bar{\boldsymbol{E}}(\boldsymbol{r}, t)e^{-i\omega_{pe}t} + \bar{\boldsymbol{E}}^*(\boldsymbol{r}, t)e^{i\omega_{pe}t}\} \\ &= \tfrac{1}{2}\bar{\boldsymbol{E}}(\boldsymbol{r}, t)e^{-i\omega_{pe}t} + \text{c.c.},\end{aligned} \tag{6.25}$$

where c.c. denotes the complex conjugate. When we take the second partial time derivative of $\tilde{\boldsymbol{E}}$, expressed in the form of eqn (6.25), we obtain

$$\frac{\partial^2 \tilde{\boldsymbol{E}}}{\partial t^2} = \frac{1}{2}\left(-\omega_{pe}^2 \bar{\boldsymbol{E}} - 2i\omega_{pe}\frac{\partial \bar{\boldsymbol{E}}}{\partial t} + \frac{\partial^2 \bar{\boldsymbol{E}}}{\partial t^2}\right)e^{-i\omega_{pe}t} + \text{c.c.} \tag{6.26}$$

Now $\bar{\boldsymbol{E}}(\boldsymbol{r}, t)$ is by definition a slowly varying amplitude. This means that

$$|\omega_{pe}^2 \bar{\boldsymbol{E}}| \gg \left|\omega_{pe}\frac{\partial \bar{\boldsymbol{E}}}{\partial t}\right| \gg \left|\frac{\partial^2 \bar{\boldsymbol{E}}}{\partial t^2}\right| \tag{6.27}$$

so that we may safely neglect the contribution of $\partial^2 \bar{\boldsymbol{E}}/\partial t^2$ in eqn (6.26). Using this result, we now substitute the expression for $\tilde{\boldsymbol{E}}$ given by eqn (6.25) into eqn (6.21). We obtain

$$-2i\omega_{pe}\frac{\partial \bar{\boldsymbol{E}}}{\partial t} - 3v_{Te}^2 \nabla(\nabla \cdot \bar{\boldsymbol{E}}) + c^2 \nabla \times (\nabla \times \bar{\boldsymbol{E}}) + \omega_{pe}^2 \frac{\delta n}{n_0}\bar{\boldsymbol{E}} = 0. \tag{6.28}$$

Here we have cancelled the common factor $e^{-i\omega_{pe}t}$; for simplicity we have dropped the '+c.c.' term, since the vanishing of the complex conjugate of eqn (6.28) is implicit in the equation itself. Equation (6.28) is the first Zakharov equation. It is a non-linear evolution equation for the slowly varying complex amplitude $\bar{\boldsymbol{E}}(\boldsymbol{r}, t)$ of the rapidly oscillating electric field, which displays the coupling of $\bar{\boldsymbol{E}}$ to the slowly varying local number density δn. Equation (6.28) differs from eqn (6.21) in three significant ways. First, it is mathematically simpler than eqn (6.21), since it contains only the first partial time derivative of the new variable $\bar{\boldsymbol{E}}$, rather than the second partial time derivative of $\tilde{\boldsymbol{E}}$. Second, eqn (6.28) contains only slowly varying quantities, whereas $\tilde{\boldsymbol{E}}$ in eqn (6.21) includes rapid oscillation—indeed, as we have seen, eqn (6.21) can be used to derive the Langmuir dispersion relation eqn (6.22). Third, eqn (6.28) is a complex rather than a real expression. The fact that it has two components, real and imaginary, is simply a reflection of a basic degree of freedom of the electric field, which has independent amplitudes of oscillation as $\cos \omega_{pe}t$ and $\sin \omega_{pe}t$. All these changes between eqn (6.21) and eqn (6.28) are consequences of writing $\tilde{\boldsymbol{E}}(\boldsymbol{r}, t)$ in the form given in eqn (6.25).

We can now consider the slow-timescale components of the fluid equations of motion. Let us return to eqn (6.18), substitute for n_e, $\boldsymbol{E}$, and $\boldsymbol{v}_e$ from eqns (6.2), (6.6), and (6.9) respectively, and average the resulting equation over many rapid oscillations. The terms that are linear in fast-timescale quantities average to zero, and we are left with

$$\frac{\partial}{\partial t}\delta\boldsymbol{v}_{\mathrm{e}}+\langle(\tilde{\boldsymbol{v}}_{\mathrm{e}}\cdot\boldsymbol{\nabla})\tilde{\boldsymbol{v}}_{\mathrm{e}}\rangle=-\frac{e}{m}\boldsymbol{E}_{\mathrm{slow}}-\frac{k_{\mathrm{B}}T_{\mathrm{e}}}{mn_0}\boldsymbol{\nabla}\,\delta n_{\mathrm{e}}. \tag{6.29}$$

Recall from our discussion after eqn (6.19) that we take $\gamma_{\mathrm{e}}=1$ here, because the slow-timescale perturbation is isothermal for the electrons. In eqn (6.29), $\partial\delta\boldsymbol{v}_{\mathrm{e}}/\partial t$ is the partial time derivative of a slow timescale quantity, and necessarily $|\delta\boldsymbol{v}_{\mathrm{e}}|\ll|\tilde{\boldsymbol{v}}_{\mathrm{e}}|$. We can therefore safely neglect this term. In dealing with the term $\langle(\tilde{\boldsymbol{v}}_{\mathrm{e}}\cdot\boldsymbol{\nabla})\tilde{\boldsymbol{v}}_{\mathrm{e}}\rangle$, we require first of all a vector identity:

$$(\tilde{\boldsymbol{v}}_{\mathrm{e}}\cdot\boldsymbol{\nabla})\tilde{\boldsymbol{v}}_{\mathrm{e}}=\tfrac{1}{2}\boldsymbol{\nabla}(\tilde{\boldsymbol{v}}_{\mathrm{e}}\cdot\tilde{\boldsymbol{v}}_{\mathrm{e}})+\tilde{\boldsymbol{v}}_{\mathrm{e}}\times(\boldsymbol{\nabla}\times\tilde{\boldsymbol{v}}_{\mathrm{e}}). \tag{6.30}$$

In general, the term $\boldsymbol{\nabla}\times\tilde{\boldsymbol{v}}_{\mathrm{e}}$ is very small because the fast-timescale waves are dominated by an electrostatic component with $\tilde{\boldsymbol{E}}$, $\tilde{\boldsymbol{v}}_{\mathrm{e}}$, and the wavenumber $\boldsymbol{k}$ all directed along the same axis, so that $\boldsymbol{\nabla}\times\tilde{\boldsymbol{v}}_{\mathrm{e}}\sim\boldsymbol{k}\times\tilde{\boldsymbol{v}}_{\mathrm{e}}$ is small. We therefore neglect this term in eqn (6.30), and eqn (6.29) becomes

$$\tfrac{1}{2}\langle\boldsymbol{\nabla}(\tilde{\boldsymbol{v}}_{\mathrm{e}}\cdot\tilde{\boldsymbol{v}}_{\mathrm{e}})\rangle=-\frac{e}{m}\boldsymbol{E}_{\mathrm{slow}}-\frac{k_{\mathrm{B}}T_{\mathrm{e}}}{mn_0}\boldsymbol{\nabla}\,\delta n_{\mathrm{e}}. \tag{6.31}$$

Our next aim, following our usual strategy, is to replace fluid variables by electric field variables. From eqn (6.19), we have to leading order $\partial\tilde{\boldsymbol{v}}_{\mathrm{e}}/\partial t=-(e/m)\tilde{\boldsymbol{E}}$. We know that when it operates on a fast-timescale variable, $\partial/\partial t$ brings down a factor ω, and by eqn (6.22) $\omega\simeq\omega_{\mathrm{pe}}$. Hence eqn (6.19) gives to leading order

$$\tilde{\boldsymbol{v}}_{\mathrm{e}}\cdot\tilde{\boldsymbol{v}}_{\mathrm{e}}=\left(\frac{e}{m\omega_{\mathrm{pe}}}\right)^2\tilde{\boldsymbol{E}}\cdot\tilde{\boldsymbol{E}}, \tag{6.32}$$

and we shall use this in eqn (6.31). First, however, we need to carry out the averaging process denoted by $\langle\;\rangle$ in eqn (6.31). It follows from eqn (6.25) that we can write

$$\tilde{\boldsymbol{E}}\cdot\tilde{\boldsymbol{E}}=\tfrac{1}{2}\bar{\boldsymbol{E}}\cdot\bar{\boldsymbol{E}}^*+\tfrac{1}{4}\bar{\boldsymbol{E}}\cdot\bar{\boldsymbol{E}}\mathrm{e}^{-2\mathrm{i}\omega_{\mathrm{pe}}t}+\tfrac{1}{4}\bar{\boldsymbol{E}}^*\cdot\bar{\boldsymbol{E}}^*\mathrm{e}^{2\mathrm{i}\omega_{\mathrm{pe}}t}. \tag{6.33}$$

The second and third terms on the right-hand side are rapidly oscillating, and average to zero. It follows that

$$\langle\tilde{\boldsymbol{E}}\cdot\tilde{\boldsymbol{E}}\rangle=\tfrac{1}{2}\bar{\boldsymbol{E}}\cdot\bar{\boldsymbol{E}}^*=\tfrac{1}{2}\,|\bar{\boldsymbol{E}}|^2. \tag{6.34}$$

Using eqns (6.32) and (6.34), eqn (6.31) becomes

$$\frac{e^2}{4m^2\omega_{\mathrm{pe}}^2}\boldsymbol{\nabla}\,|\bar{\boldsymbol{E}}|^2=-\frac{e}{m}\boldsymbol{E}_{\mathrm{slow}}-\frac{k_{\mathrm{B}}T_{\mathrm{e}}}{mn_0}\boldsymbol{\nabla}\,\delta n_{\mathrm{e}}. \tag{6.35}$$

The non-linear term $\boldsymbol{\nabla}\,|\bar{\boldsymbol{E}}(\boldsymbol{r},t)|^2$ in eqn (6.35) thus represents the slow-timescale average of the fast-timescale non-linearity $(\tilde{\boldsymbol{v}}_{\mathrm{e}}\cdot\boldsymbol{\nabla})\tilde{\boldsymbol{v}}_{\mathrm{e}}$ in the equation of motion for the electron fluid.

Now we consider, for the first time, the equation of motion for the ion fluid:

$$\frac{\partial}{\partial t}\delta\boldsymbol{v}_{\mathrm{i}}+(\delta\boldsymbol{v}_{\mathrm{i}}\cdot\boldsymbol{\nabla})\,\delta\boldsymbol{v}_{\mathrm{i}}=\frac{e}{M}\boldsymbol{E}_{\mathrm{slow}}-\frac{3k_{\mathrm{B}}T_{\mathrm{i}}}{Mn_0}\boldsymbol{\nabla}\,\delta n_{\mathrm{i}}. \tag{6.36}$$

Here, we have taken $\gamma_{\mathrm{i}}=3$; this reflects the fact that the slow-timescale perturbation is adiabatic for the ions although, as has been said, it is isothermal for the more mobile electrons. In eqn (6.36), we first neglect the slow-timescale ion non-linearity since $\delta\boldsymbol{v}_{\mathrm{i}}$ is small compared to $\tilde{\boldsymbol{v}}_{\mathrm{e}}$, whose slow-timescale non-linearity has already been included. Next, we substitute for $\boldsymbol{E}_{\mathrm{slow}}$ in eqn (6.36) using eqn (6.35). We obtain

$$\frac{\partial}{\partial t}\delta\boldsymbol{v}_{\mathrm{i}}=-\frac{e^2}{4mM\omega_{\mathrm{pe}}^2}\boldsymbol{\nabla}\,|\bar{\boldsymbol{E}}|^2-\frac{k_{\mathrm{B}}T_{\mathrm{e}}}{Mn_0}\boldsymbol{\nabla}\,\delta n_{\mathrm{e}}-\frac{3k_{\mathrm{B}}T_{\mathrm{i}}}{Mn_0}\boldsymbol{\nabla}\,\delta n_{\mathrm{i}}. \tag{6.37}$$

The continuity equation for the ion fluid is

$$\frac{\partial}{\partial t}\delta n_{\mathrm{i}}+\boldsymbol{\nabla}\cdot\{(n_0+\delta n_{\mathrm{i}})\,\delta\boldsymbol{v}_{\mathrm{i}}\}=0, \tag{6.38}$$

and neglecting the slow-timescale non-linearity $\delta n_{\mathrm{i}}\,\delta\boldsymbol{v}_{\mathrm{i}}$, this gives

$$\frac{\partial}{\partial t}\delta n_{\mathrm{i}}+n_0\boldsymbol{\nabla}\cdot\delta\boldsymbol{v}_{\mathrm{i}}=0. \tag{6.39}$$

Now the effect of slow-timescale electrical non-neutrality has already been included through $\boldsymbol{E}_{\mathrm{slow}}$. In view of eqn (6.4), there is no need to distinguish between δn_{e} and δn_{i} in any of the other terms, so that we shall write δn for both variables in eqns (6.37) and (6.39). Taking $\partial/\partial t$ of eqn (6.39), and substituting for $\partial\delta\boldsymbol{v}_{\mathrm{i}}/\partial t$ using eqn (6.37), we have

$$\frac{\partial^2}{\partial t^2}\delta n-\frac{k_{\mathrm{B}}(T_{\mathrm{e}}+3T_{\mathrm{i}})}{M}\nabla^2\,\delta n-\frac{n_0e^2}{4mM\omega_{\mathrm{pe}}^2}\nabla^2\,|\bar{\boldsymbol{E}}|^2=0. \tag{6.40}$$

Defining the ion-acoustic velocity c_{s} (compare V_{s} defined by eqn (4.64)) by

$$c_{\mathrm{s}}=\surd(k_{\mathrm{B}}(T_{\mathrm{e}}+3T_{\mathrm{i}})/M), \tag{6.41}$$

and recalling that $\omega_{\mathrm{pe}}^2=n_0e^2/m\varepsilon_0$ from eqn (1.6), we can write eqn (6.40) as

$$\left(\frac{\partial^2}{\partial t^2}-c_{\mathrm{s}}^2\nabla^2\right)\delta n=\frac{\varepsilon_0}{4M}\nabla^2\,|\bar{\boldsymbol{E}}|^2. \tag{6.42}$$

This is the second Zakharov equation. Like the first, eqn (6.28), it involves only δn and $\bar{\boldsymbol{E}}$, so that we now have two equations with two variables—a closed system that must be solved self-consistently. The nonlinear term proportional to $\nabla^2 |\bar{\boldsymbol{E}}|^2$ is referred to as the ponderomotive—that is, 'mass-moving'—force. It originates in the term $\langle(\tilde{\boldsymbol{v}}_e \cdot \nabla)\tilde{\boldsymbol{v}}_e\rangle$ in eqn (6.29). We therefore need to consider how it is that the slow-timescale average of a non-linear term which is produced by the fast-timescale motion of the light electrons can affect the dynamics of the much heavier ions. The answer rests on the fact that while the masses of the electrons and ions are widely different, the magnitude of their charges is the same, so that a given electric field produces a force of equal magnitude on the electron and ion fluids. First, because the electrons are light, the magnitude of the electron velocity $|\tilde{\boldsymbol{v}}_e| \simeq |e\tilde{\boldsymbol{E}}/m\omega_{pe}|$ is large. As a result, $\langle(\tilde{\boldsymbol{v}}_e \cdot \nabla)\tilde{\boldsymbol{v}}_e\rangle$ dominates the left-hand side of eqn (6.29). We then arrive at eqn (6.31), which expresses the slow-timescale balance between this non-linear term, the electron pressure term, and the field $\boldsymbol{E}_{slow}$ which is associated with slow-timescale electrical charge non-neutrality. Thus eqn (6.31) can be interpreted as relating $\boldsymbol{E}_{slow}$ to its two sources, namely the non-linear term and the electron pressure term. Now $\boldsymbol{E}_{slow}$ also acts on the ion fluid with a force equal in magnitude $|e\boldsymbol{E}_{slow}|$ to that with which it acts on the electrons. This force is shown in eqn (6.36). Its replacement by the expression for its source—eqn (6.31) or equivalently eqn (6.35)—brings the averaged fast-timescale electron non-linearity into the equation of motion for the ion fluid. Alternatively, but a little more loosely, we could move the term $\langle(\tilde{\boldsymbol{v}}_e \cdot \nabla)\tilde{\boldsymbol{v}}_e\rangle$ onto the right-hand side of eqn (6.29), and interpret it as a third independent force that contributes to the small quantity $\partial\delta\boldsymbol{v}_e/\partial t$. In this case, we may say that any resulting change in the time-evolution of $\delta\boldsymbol{v}_e$ will be repeated by $\delta\boldsymbol{v}_i$, as the strong Coulomb force drags the ions along with the electrons, preventing the creation of large deviations from charge neutrality.

Finally, we recall that when we linearized eqn (6.21), we obtained the dispersion relation eqn (6.22) for Langmuir waves. Here, let us linearize eqn (6.42). Then the slow-timescale combined ion and electron density perturbations that vary as $\exp(\mathrm{i}\boldsymbol{k} \cdot \boldsymbol{r} - \mathrm{i}\omega t)$ satisfy the dispersion relation

$$\omega^2 = c_s^2 k^2. \tag{6.43}$$

These are ion-acoustic waves; we did not meet these in Chapter 3, because we concentrated on waves in cold plasma—ion-acoustic waves require the temperature to be non-zero. It follows that Langmuir waves and ion-acoustic waves are the fast-timescale and slow-timescale normal modes of an electron–ion plasma, provided that their amplitudes are sufficiently small that the non-linear coupling terms in eqns (6.21) and (6.42) are negligible. In the following sections, we shall consider the larger-amplitude cases where the non-linear coupling is not negligible, and examine some further concepts of non-linear plasma physics. To begin with, we shall need to quantify some of the remarks that we have just made.

6.2 The turbulence parameter and modulational instability

In the previous section, we obtained the non-linear equations, eqns (6.28) and (6.42), that describe the coupled motion of the intermingled electron and ion fluids that comprise the plasma. These equations both contain a single non-linear coupling term. The remaining linear terms give rise to the linear dispersion relations for Langmuir and ion-acoustic waves. It is now possible to address the fundamental question that immediately arises: under what circumstances does non-linear coupling significantly affect the dynamics of the electron and ion fluids? Equivalently, how large does the amplitude of a wave have to be in order to produce significant deviations from the linear dispersion relations given in eqns (6.22) and (6.43)?

We note first that the motion of electrons in a plasma has two distinct aspects. There is the random thermal motion, which gives rise to the pressure term in eqn (6.18). In addition, there may be coherent collective motion associated with electromagnetic or electrostatic waves, which can exist in a plasma even in the absence of any thermal motion. It is useful to define a quantity which is the ratio of the average electric field energy density associated with the rapid wave motion to the thermal energy density associated with the random motion of the electrons. This is the turbulence parameter,

$$\bar{W} = \frac{\varepsilon_0 \langle |\tilde{\boldsymbol{E}}|^2 \rangle}{2} \frac{1}{n_0 k_\mathrm{B} T_\mathrm{e}} = \frac{\varepsilon_0 \, |\bar{\boldsymbol{E}}|^2}{4 n_0 k_\mathrm{B} T_\mathrm{e}}, \tag{6.44}$$

where we have used eqn (6.34). The questions that we have raised in the first paragraph can be answered in terms of the value of this dimensionless parameter. For example, from eqns (6.19) and (6.22), to leading order

$$\langle |\tilde{\boldsymbol{v}}_\mathrm{e}|^2 \rangle = \left(\frac{e}{m \omega_\mathrm{pe}} \right)^2 \langle |\tilde{\boldsymbol{E}}|^2 \rangle = 2 \bar{W} v_\mathrm{Te}^2. \tag{6.45}$$

Here, we have used the standard definitions $\omega_\mathrm{pe}^2 = n_0 e^2 / m \varepsilon_0$ and $v_\mathrm{Te}^2 = k_\mathrm{B} T_\mathrm{e} / m$, together with eqn (6.44). Thus, the average velocity acquired by the electron in response to the fast-timescale oscillatory field $\tilde{\boldsymbol{E}}$ is small compared to its thermal velocity provided

$$\bar{W}^{\frac{1}{2}} \ll 1. \tag{6.46}$$

Under this condition, eqn (6.22) will represent the dispersion relation for electrostatic waves very accurately—we note that electron thermal motion itself produces only a small correction to $\omega = \omega_\mathrm{pe}$ in the range of interest $k \leqslant k_\mathrm{D}$ where Landau damping is small. Now let us turn to the left-hand side of eqn (6.18), and consider the ratio of the average of the non-linear term to the linear term. For the non-linear term, using eqns

(6.30) and (6.45), we have

$$\langle(\tilde{\boldsymbol{v}}_e \cdot \boldsymbol{\nabla})\tilde{\boldsymbol{v}}_e\rangle = v_{Te}^2 \boldsymbol{\nabla}\bar{W}. \tag{6.47}$$

For the linear term, we shall be interested in the square root of

$$\left\langle\left|\frac{\partial \tilde{\boldsymbol{v}}_e}{\partial t}\right|^2\right\rangle \simeq \omega_{pe}^2\langle|\tilde{\boldsymbol{v}}_e|^2\rangle = 2\omega_{pe}^2 \bar{W} v_{Te}^2, \tag{6.48}$$

where again we have used eqn (6.45). Combining eqns (6.47) and (6.48), the ratio of interest is

$$\frac{\langle|(\tilde{\boldsymbol{v}}_e \cdot \boldsymbol{\nabla})\tilde{\boldsymbol{v}}_e|\rangle}{\sqrt{[\langle|\partial\tilde{\boldsymbol{v}}_e/\partial t|^2\rangle]}} = \frac{1}{\sqrt{2}}\,\frac{|\boldsymbol{\nabla}\bar{W}|}{k_D \bar{W}^{\frac{1}{2}}}, \tag{6.49}$$

where we have used the standard definition $k_D = \omega_{pe}/v_{Te}$ of the Debye wavenumber. Suppose that the spatial variation of $\bar{W}$—which by eqn (6.44) reflects the spatial variation of the electric field $\tilde{\boldsymbol{E}}$—has characteristic wavenumber k, so that $|\boldsymbol{\nabla}\bar{W}| \simeq k\bar{W}$. Then it follows from eqn (6.49) that the non-linear term in the fast-timescale equation of motion for the electron fluid is small compared to the linear term provided that

$$\bar{W}^{\frac{1}{2}} \frac{k}{k_D} \ll 1. \tag{6.50}$$

So far in this section, we have considered the turbulence parameter $\bar{W}$ and the ratio k/k_D from the point of view of electron dynamics. Now let us turn to the ion dynamics, which are governed by eqn (6.42). Dividing eqn (6.42) by the equilibrium number density n_0, we obtain

$$\left(\frac{\partial^2}{\partial t^2} - c_s^2\nabla^2\right)\frac{\delta n}{n_0} = \frac{\varepsilon_0}{4n_0 M}\nabla^2\,|\tilde{\boldsymbol{E}}|^2. \tag{6.51}$$

Now suppose that, as is often the case, the ions are significantly cooler than the electrons, so that following eqn (6.41) the restoring force for ion-acoustic waves is provided by the pressure of the hot electrons on the ions. Then by eqn (6.41), $Mc_s^2 \simeq k_B T_e$ and using eqn (6.44) we may write eqn (6.51) in the form

$$\frac{\partial^2}{\partial t^2}\frac{\delta n}{n_0} = c_s^2\nabla^2\left(\frac{\delta n}{n_0} + \bar{W}\right). \tag{6.52}$$

Equation (6.52) has two immediate implications. First, the ion dynamics are only negligibly affected by the non-linearity associated with fast-timescale electron oscillations provided that

$$\bar{W} \ll \left|\frac{\delta n}{n_0}\right|, \tag{6.53}$$

where by definition $|\delta n/n_0| \leqslant 1$. Second, we note that $\partial^2(\delta n/n_0)/\partial t^2$ is zero when

$$\frac{\delta n}{n_0} = -\bar{W}. \tag{6.54}$$

This is the quasistatic solution of the second Zakharov equation. It represents a state in which regions of enhanced electric field amplitude correspond to regions of greater depletion of the ion number density.

Now let us consider ion dynamics in the context of the expression for the evolution of the electric field $\tilde{\boldsymbol{E}}$ given by eqn (6.21). Neglecting as usual the term $\nabla \times (\nabla \times \tilde{\boldsymbol{E}})$, we seek solutions where $\tilde{\boldsymbol{E}}$ varies as $\exp(\mathrm{i}\boldsymbol{k} \cdot \boldsymbol{r} - \mathrm{i}\omega t)$. Retaining the non-linear coupling term, we obtain

$$\omega^2 = \omega_{\mathrm{pe}}^2\left(1 + \frac{3k^2}{k_{\mathrm{D}}^2} + \frac{\delta n}{n_0}\right). \tag{6.55}$$

Setting $\delta n = 0$ in eqn (6.55) yields the standard Langmuir dispersion relation eqn (6.22) as required. Let us consider cases of density depletion relative to the equilibrium level, that is

$$\delta n / n_0 < 0, \tag{6.56}$$

such as follow from the quasistatic solution eqn (6.54) of eqn (6.52). First, suppose that

$$|\delta n / n_0| < 3k^2 / k_{\mathrm{D}}^2. \tag{6.57}$$

We note that in general k^2/k_{D}^2 may have a value substantially smaller than unity. When eqns (6.56) and (6.57) apply, eqn (6.55) gives for a particular wavenumber k a frequency ω that is lower than it would be in the absence of density depletion. Conversely, if a wave has the same value for ω in a region of density depletion that it has where $\delta n = 0$, its wavenumber k must be smaller where $\delta n = 0$ than it is in the region of density depletion.

Now let us suppose that eqn (6.56) still applies, but instead of eqn (6.57) we have

$$|\delta n / n_0| > 3k^2 / k_{\mathrm{D}}^2. \tag{6.58}$$

In this case, for the wavenumber of interest, the magnitude of the density depletion term in eqn (6.55) exceeds the linear term associated with electron thermal motion. It follows from eqns (6.55), (6.56), and (6.58) that

$$\omega < \omega_{\mathrm{pe}}. \tag{6.59}$$

In many circumstances, when the approximations that govern geometric optics are applicable, the frequency of a wave remains constant as it propagates through an inhomogeneous medium, while its wavenumber changes. Now it follows from eqn (6.55) that there are no waves in the rest of the plasma, away from the region where eqns (6.56) and (6.58) apply, whose frequencies satisfy eqn (6.59). Thus, waves whose wavenumbers k are small enough to satisfy eqn (6.58) cannot propagate out of the region of density depletion: they are trapped. If the quasistatic solution given by eqn (6.54) applies, we can use eqn (6.58) to define the region where trapping becomes possible in terms of the turbulence parameter:

$$\bar{W} > 3k^2/k_{\mathrm{D}}^2. \tag{6.60}$$

It is interesting to note that the process of trapping electrostatic waves in regions of density depletion is unstable. Once started, it will grow. Suppose that in some region eqn (6.56) applies, and that accordingly eqn (6.58) is satisfied by all k less than some value k_0. We then increase $|\delta n/n_0|$ slightly. It follows from eqn (6.58) that certain waves with values of k slightly larger than k_0, that were previously free, will become trapped. The presence of these additional trapped waves will enhance the local value of $\bar{W}$, and the associated increase in the ponderomotive force will expel further ions, increasing the value of $|\delta n/n_0|$ further by eqn (6.54). This will cause the trapping of additional waves; we are clearly locked into an unstable, self-enhancing cycle. The process will terminate only when it has changed the local physical conditions so greatly that the equations that we have been using cease to be appropriate. This will occur when lengthscales of a few Debye lengths are reached, as the region of density depletion and electric field enhancement is compressed by the higher-density plasma surrounding it.

This instability is referred to as the modulational instability. It is so-called because it describes the instability of a rapidly oscillating electric field in a plasma when it is subjected to low-frequency modulation. This modulation affects the ion density through the non-linear ponderomotive force, in a manner which is described by eqn (6.42). The resulting change in the ion density couples back to the electric field, through the non-linear term in eqn (6.21) or equivalently eqn (6.28). We have seen how these couplings lead to the production of narrow cavities where the density of ions and electrons is depleted, and electric field energy is concentrated. These structures, which have been observed experimentally in laboratory plasmas, are referred to as cavitons—a hybrid word from cavity and soliton. We shall therefore close this section with a brief discussion of soliton solutions of the Zakharov equations.

First, we need to establish what is meant by a soliton. Consider a partial differential equation for some field amplitude ϕ that has two independent variables, a spatial variable x and time t. Any solution ϕ that depends on x and t only in the combination $\xi = x - ut$, where u is constant, is called a travelling wave. A pulselike travelling wave $\phi(\xi)$ whose dependence on ξ is localized to some finite range of values of ξ, and which takes a constant value independent of ξ outside this range, is called a solitary wave. A soliton is a particularly robust type of solitary wave, that emerges from collisions with other solitary waves unchanged, except possibly for a phase shift $\xi = x - ut \rightarrow x - ut + \delta$. Next, let us consider what classes of equation may have soliton solutions. First, consider the familiar wave equation

$$\frac{\partial^2 \phi}{\partial t^2} = u^2 \frac{\partial^2 \phi}{\partial x^2}. \tag{6.61}$$

In general, the solution ϕ may be built up from the sum of many components, each oscillating as $\exp(\mathrm{i}kx - \mathrm{i}\omega t)$ where k is different for each component. It follows from eqn (6.61) that

$$\omega = \pm uk. \tag{6.62}$$

Equation (6.62) is non-dispersive, that is the velocity ω/k of each field component associated with a particular wavenumber k is independent of the value of k. This ensures that a travelling wave solution of eqn (6.61) that starts out with a pulselike shape will retain it, since all of its components will travel at the same speed. Furthermore, the familiar principle of linear superposition ensures that such solitary wave solutions of eqn (6.61) will pass through each other without disruption. Hence we conclude that eqn (6.61), which is a representative linear non-dispersive equation, has soliton solutions. Now consider a relation between ω and k that is dispersive, such as eqn (6.22). In this case, ω/k depends on k, so that solitary waves are not possible. A travelling wave that started out with a pulselike shape would gradually spread out and break up, as the field components associated with different values of k move apart with different phase velocities ω/k. Thus, solitary wave solutions of linear dispersive partial differential equations are not possible. Next consider a non-linear equation without dispersion. An example can be obtained by adding a term ϕ^2 to the right-hand side of eqn (6.61), so that the non-dispersive relation eqn (6.62) would be recovered in the linearized limit $\phi^2 \to 0$. As we shall see in the next section, an effect of non-linearity is to cause energy to flow between field components with different values of k. In particular, energy may flow to components whose initial amplitude is zero. Because the equation is non-dispersive, all the components of the solution ϕ travel with the same speed, so that it is not possible to counterbalance the energy flow caused by non-linearity by a catching-up process among the modes. It follows that a nonlinear, non-dispersive partial differential equation cannot have soliton solutions. Conversely, in a non-linear dispersive equation, it may be possible to balance the non-linear and dispersive energy flows against each other, so that soliton solutions are possible.

Now let us consider again eqns (6.28) and (6.42). From eqn (6.42), we can obtain an expression for $\delta n/n_0$ in terms of $\bar{\boldsymbol{E}}$. If we substitute this expression for $\delta n/n_0$ into eqn (6.28), it will contribute a term that is non-linear in $\bar{\boldsymbol{E}}$. For example, let us use the static solution eqn (6.54) of the slow timescale equation. Then eqn (6.28) becomes

$$\mathrm{i}\omega_{\mathrm{pe}}\frac{\partial\bar{\boldsymbol{E}}}{\partial t}+\frac{3}{2}v_{\mathrm{Te}}^2\nabla(\nabla\cdot\bar{\boldsymbol{E}})+\frac{\varepsilon_0\omega_{\mathrm{pe}}^2}{8n_0k_{\mathrm{B}}T_{\mathrm{e}}}|\bar{\boldsymbol{E}}|^2\bar{\boldsymbol{E}}=0. \tag{6.63}$$

Here, we have used the definition eqn (6.44) of $\bar{W}$, and in eqn (6.28) we have neglected the term $\nabla\times(\nabla\times\bar{\boldsymbol{E}})$ as usual. Equations that have the general form of eqn (6.63) occur widely in physics, and are known generically as the non-linear Schrödinger equation. We know already that eqn (6.28) is dispersive, since $\partial\bar{\boldsymbol{E}}/\partial t$ will contribute a frequency and $\nabla(\nabla\cdot\bar{\boldsymbol{E}})\sim k^2$; this is just the thermal correction to the Langmuir dispersion relation. Thus, quite generally, eqns (6.28) and (6.42) together produce a non-linear, dispersive system. As such, it may have soliton solutions. These have the form

$$\bar{\boldsymbol{E}}=\hat{\boldsymbol{e}}_xE_0\mathrm{e}^{\mathrm{i}k_0x-\mathrm{i}\omega_0t}, \tag{6.64}$$

$$E_0\propto A\,\mathrm{sech}\,A(x-ut), \tag{6.65}$$

$$\frac{\delta n}{n_0}\propto -2A^2\,\mathrm{sech}^2\,A(x-ut). \tag{6.66}$$

Here ω_0 and k_0 are related by the dispersion relation that follows from eqn (6.28) when it is linearized: $\omega_0\sim k_0^2$, which is the thermal correction in eqn (6.22)—the leading-order frequency ω_{pe} having been removed in the transition from $\tilde{\boldsymbol{E}}$ to $\bar{\boldsymbol{E}}$ in eqn (6.25). The solution for E_0 is referred to as an envelope soliton, because it describes the amplitude of the rapidly varying electric field, rather than the electric field itself. We note that the

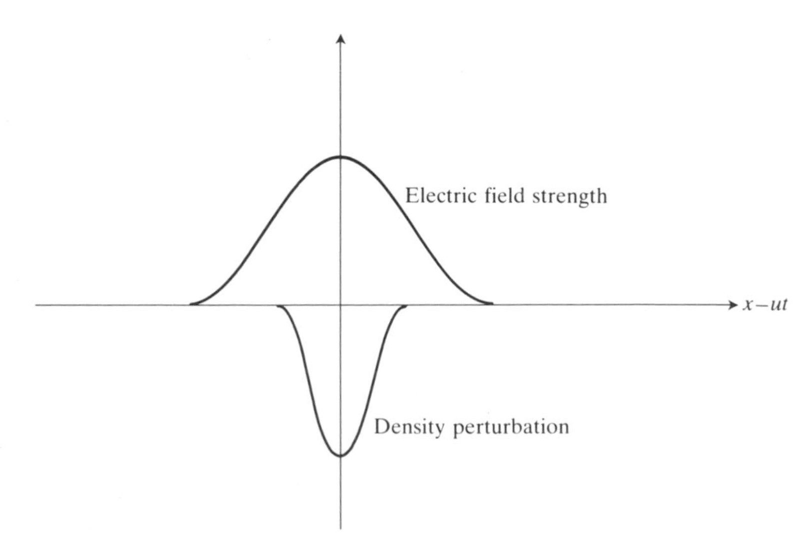

Fig. 6.1 Soliton solutions.

soliton solutions describe regions of electric field concentration and density depletion, see Fig. 6.1. These correspond to the cavities whose production by the modulational instability was described above—hence the term caviton. Thus, two aspects of the Zakharov equations—the structures produced by the instability that they describe, and their soliton solutions—appear to fit together. However, there remains a problem. The solutions eqns (6.64) to (6.66) are independent of the spatial coordinates y and z, so that the solitons are planar. It can be shown that these solitons are actually unstable to transverse perturbations. Thus, eqns (6.64) to (6.66) cannot represent the end-point of the evolution of the plasma described by the Zakharov equations. It has not proved possible, as was once hoped, to represent the plasma in terms of an ensemble of solitons. Instead, the question of how best to describe the two-fluid plasma, evolving under the Zakharov equations, remains open.

6.3 Non-linear wave coupling

We shall now examine the way in which the normal modes of linear theory are affected by the non-linear coupling between electric field and density perturbations that is described by eqns (6.21) and (6.42). First, we construct a single electric field perturbation whose wavevector is $\boldsymbol{k}_0$. From eqn (6.22), the corresponding frequency ω_0 of the linear normal mode satisfies

$$\omega_0^2 = \omega_{pe}^2(1 + 3k_0^2/k_D^2). \tag{6.67}$$

In linear theory, the amplitude of this wave is constant. However, non-linear coupling will cause the amplitude to evolve in space and time. We therefore write this wave in a form which allows for a slowly varying amplitude, following the approach of eqn (6.25):

$$\tilde{\boldsymbol{E}}_0(\boldsymbol{r}, t) = \tfrac{1}{2}\bar{\boldsymbol{E}}_0(\boldsymbol{r}, t)\mathrm{e}^{\mathrm{i}\boldsymbol{k}_0\cdot\boldsymbol{r} - \mathrm{i}\omega_0 t} + \text{c.c.}, \tag{6.68}$$

where

$$\left|\frac{\partial \bar{\boldsymbol{E}}_0}{\partial t}\right| \ll \omega_0\,|\bar{\boldsymbol{E}}_0|, \qquad |\boldsymbol{\nabla}\bar{\boldsymbol{E}}_0| \ll k_0\,|\bar{\boldsymbol{E}}_0|. \tag{6.69}$$

Next, we adopt a similar approach to density perturbations. Because these are low-frequency perturbations, we do not need to split their time-dependence into two parts, as was necessary for the high-frequency wave eqn (6.68). We write the density perturbation with wavevector $\boldsymbol{q}$ in the form

$$\delta n(\boldsymbol{r}, t) = \tfrac{1}{2}n_q(\boldsymbol{r}, t)\mathrm{e}^{\mathrm{i}\boldsymbol{q}\cdot\boldsymbol{r}} + \text{c.c.}, \tag{6.70}$$

where

$$|\boldsymbol{\nabla} n_q| \ll q n_q. \tag{6.71}$$

From eqn (6.42), we know that in the linear limit, $n_q(\boldsymbol{r}, t)$ reduces to a constant amplitude times $\exp(-ic_s qt)$, corresponding to an ion acoustic wave.

Now let us calculate the non-linear term that arises from the coupling of these two perturbations. In eqn (6.21), we require the term

$$\delta n\tilde{\boldsymbol{E}} = \tfrac{1}{4}(n_q\bar{\boldsymbol{E}}_0 e^{i(\boldsymbol{k}_0+\boldsymbol{q})\cdot\boldsymbol{r}} + n_q^*\bar{\boldsymbol{E}}_0 e^{i(\boldsymbol{k}_0-\boldsymbol{q})\cdot\boldsymbol{r}})e^{-i\omega_0 t} + \text{c.c.}, \tag{6.72}$$

where the expression on the right is obtained using eqns (6.68) and (6.70). This term drives the non-linear evolution of the electric field. It is characterized by two wavevectors, $\boldsymbol{k}_0 \pm \boldsymbol{q}$, which differ from the wavevector $\boldsymbol{k}$ of the initial electric field perturbation. For a consistent treatment, we must therefore include two additional waves in $\tilde{\boldsymbol{E}}$, whose amplitudes are initially much smaller than $|\bar{\boldsymbol{E}}_0|$. Following eqn (6.68), we write these as

$$\tilde{\boldsymbol{E}}_1(\boldsymbol{r}, t) = \tfrac{1}{2}\bar{\boldsymbol{E}}_1(\boldsymbol{r}, t)e^{i(\boldsymbol{k}_0-\boldsymbol{q})\cdot\boldsymbol{r}-i\omega_1 t} + \text{c.c.}, \tag{6.73}$$

$$\tilde{\boldsymbol{E}}_2(\boldsymbol{r}, t) = \tfrac{1}{2}\bar{\boldsymbol{E}}_2(\boldsymbol{r}, t)e^{i(\boldsymbol{k}_0+\boldsymbol{q})\cdot\boldsymbol{r}-i\omega_2 t} + \text{c.c.} \tag{6.74}$$

Here, ω_1 and ω_2 are the frequencies that follow from the linear dispersion relation eqn (6.22) for the corresponding wavevectors. They satisfy

$$\omega_1^2 = \omega_{pe}^2(1 + 3\,|\boldsymbol{k}_0 - \boldsymbol{q}|^2/k_D^2), \tag{6.75}$$

$$\omega_2^2 = \omega_{pe}^2(1 + 3\,|\boldsymbol{k}_0 + \boldsymbol{q}|^2/k_D^2). \tag{6.76}$$

The amplitudes $\bar{\boldsymbol{E}}_1$ and $\bar{\boldsymbol{E}}_2$ are slowly varying in the sense defined by eqn (6.69). Thus, for example, we shall neglect $\partial^2\bar{\boldsymbol{E}}_1/\partial t^2$ compared to $\omega_1\,\partial\bar{\boldsymbol{E}}_1/\partial t$. Using eqns (6.72) and (6.73) in eqn (6.21), we obtain the evolution equation for the component of the electric field whose wavevector is $\boldsymbol{k}_0 - \boldsymbol{q}$ by equating coefficients of $\exp\{i(\boldsymbol{k}_0 - \boldsymbol{q})\cdot\boldsymbol{r}\}$:

$$\frac{1}{2}\left(-\omega_1^2\bar{\boldsymbol{E}}_1 - 2i\omega_1\frac{\partial\bar{\boldsymbol{E}}_1}{\partial t} + \omega_{pe}^2\bar{\boldsymbol{E}}_1 + 3v_{Te}^2\,|\boldsymbol{k}_0 - \boldsymbol{q}|^2\bar{\boldsymbol{E}}_1 + \frac{\omega_{pe}^2}{2}\frac{n_q^*}{n_0}\bar{\boldsymbol{E}}_0 e^{-i(\omega_0-\omega_1)t}\right)e^{i(\boldsymbol{k}_0-\boldsymbol{q})\cdot\boldsymbol{r}}e^{-i\omega_1 t} + \text{c.c.} = 0. \tag{6.77}$$

Here, as usual, we have neglected the term involving $\nabla\times(\nabla\times\tilde{\boldsymbol{E}})$ because the wave is predominantly electrostatic. Using eqn (6.75), eqn (6.77) yields

$$i\frac{\partial\bar{\boldsymbol{E}}_1}{\partial t} = \frac{\omega_{pe}^2}{4\omega_1}\frac{n_q^*}{n_0}\bar{\boldsymbol{E}}_0 e^{-i(\omega_0-\omega_1)t}. \tag{6.78}$$

This shows how the non-linear coupling of the initial electric field perturbation eqn (6.68) to the density perturbation eqn (6.70) causes the amplitude of the wave defined at (6.73) to evolve in time. We now use the same approach for the component of the electric field oscillation whose wavenumber is $\boldsymbol{k}_0 + \boldsymbol{q}$. From eqns (6.21), (6.72), and (6.74), we obtain

$$\frac{1}{2}\Bigg(-\omega_2^2\bar{\boldsymbol{E}}_2 - 2\mathrm{i}\omega_2\frac{\partial\bar{\boldsymbol{E}}_2}{\partial t} + \omega_{\mathrm{pe}}^2\bar{\boldsymbol{E}}_2 + 3v_{\mathrm{Te}}^2\,|\boldsymbol{k}_0 + \boldsymbol{q}|^2\bar{\boldsymbol{E}}_2 + \frac{\omega_{\mathrm{pe}}^2}{2}\frac{n_q}{n_0}\bar{\boldsymbol{E}}_0\mathrm{e}^{-\mathrm{i}(\omega_0-\omega_2)t}\Bigg)\mathrm{e}^{\mathrm{i}(\boldsymbol{k}_0+\boldsymbol{q})\cdot\boldsymbol{r}}\mathrm{e}^{-\mathrm{i}\omega_2 t} + \text{c.c.} = 0. \qquad (6.79)$$

Using eqn (6.76), eqn (6.79) yields

$$\mathrm{i}\frac{\partial\bar{\boldsymbol{E}}_2}{\partial t} = \frac{\omega_{\mathrm{pe}}^2}{4\omega_2}\frac{n_q}{n_0}\bar{\boldsymbol{E}}_0\mathrm{e}^{-\mathrm{i}(\omega_0-\omega_2)t}. \qquad (6.80)$$

As usual, we consider electrostatic waves whose wavelength is long compared to the Debye length. Applying the binomial theorem to the square roots of eqns (6.67), (6.75), and (6.76), and truncating accordingly, we define

$$\delta_1 = \omega_0 - \omega_1 = \frac{3(2\boldsymbol{k}_0\cdot\boldsymbol{q} - q^2)}{2k_{\mathrm{D}}^2}\,\omega_{\mathrm{pe}}, \qquad (6.81)$$

$$\delta_2 = \omega_0 - \omega_2 = \frac{-3(2\boldsymbol{k}_0\cdot\boldsymbol{q} + q^2)}{2k_{\mathrm{D}}^2}\,\omega_{\mathrm{pe}}, \qquad (6.82)$$

$$\mu = -(\delta_1 + \delta_2)/2 = \frac{3q^2}{2k_{\mathrm{D}}^2}\,\omega_{\mathrm{pe}}. \qquad (6.83)$$

The small differences between the constant coefficients in eqns (6.78) and (6.80) can be neglected. For these purposes,

$$\frac{\omega_{\mathrm{pe}}^2}{\omega_1} = \frac{\omega_{\mathrm{pe}}^2}{\omega_2} = \frac{\omega_{\mathrm{pe}}^2}{\omega_0} = \omega_{\mathrm{pe}}. \qquad (6.84)$$

Thus, we may write eqns (6.78) and (6.80) in the form

$$\mathrm{i}\frac{\partial\bar{\boldsymbol{E}}_1}{\partial t} = \frac{\omega_{\mathrm{pe}}}{4}\frac{n_q^*}{n_0}\bar{\boldsymbol{E}}_0\mathrm{e}^{-\mathrm{i}\delta_1 t}, \qquad (6.85)$$

$$\mathrm{i}\frac{\partial\bar{\boldsymbol{E}}_2}{\partial t} = \frac{\omega_{\mathrm{pe}}}{4}\frac{n_q}{n_0}\bar{\boldsymbol{E}}_0\mathrm{e}^{-\mathrm{i}\delta_2 t}. \qquad (6.86)$$

So far, we have seen that non-linear coupling between the electrostatic wave $\tilde{\boldsymbol{E}}_0$ and the density perturbation δn generates electrostatic waves with wavevectors $\boldsymbol{k}_0 \pm \boldsymbol{q}$. Let us now turn to the second Zakharov equation, eqn (6.42), which describes how non-linear interaction between the three electrostatic waves affects the density perturbation. This involves the quantity $|\bar{\boldsymbol{E}}|^2$, which is the slow-timescale average of the square of the electric field amplitude. From eqns (6.68), (6.73), and

(6.74), we have to leading order

$$|\bar{\boldsymbol{E}}|^2 = \frac{1}{4}\big(|\bar{\boldsymbol{E}}_0|^2 + \bar{\boldsymbol{E}}_0 \cdot \bar{\boldsymbol{E}}_1^* \mathrm{e}^{\mathrm{i}\boldsymbol{q}\cdot\boldsymbol{r}-\mathrm{i}\delta_1 t} + \bar{\boldsymbol{E}}_0 \cdot \bar{\boldsymbol{E}}_2^* \mathrm{e}^{-\mathrm{i}\boldsymbol{q}\cdot\boldsymbol{r}-\mathrm{i}\delta_2 t}$$
$$+ \bar{\boldsymbol{E}}_0^* \cdot \bar{\boldsymbol{E}}_1 \mathrm{e}^{-\mathrm{i}\boldsymbol{q}\cdot\boldsymbol{r}+\mathrm{i}\delta_1 t} + \bar{\boldsymbol{E}}_0^* \cdot \bar{\boldsymbol{E}}_2 \mathrm{e}^{\mathrm{i}\boldsymbol{q}\cdot\boldsymbol{r}+\mathrm{i}\delta_2 t}\big). \quad (6.87)$$

Here, terms such as $\bar{\boldsymbol{E}}_1^* \cdot \bar{\boldsymbol{E}}_2 \exp\{2\mathrm{i}\boldsymbol{q}\cdot\boldsymbol{r} + \mathrm{i}(\delta_2 - \delta_1)t\}$ have been neglected because $|\bar{\boldsymbol{E}}_{1,2}| \ll |\bar{\boldsymbol{E}}_0|$. When such terms are retained, they generate further density perturbations with wavevectors $\pm 2\boldsymbol{q}$. All other terms that involve the product of wave field amplitudes, such as $\bar{\boldsymbol{E}}_0 \cdot \bar{\boldsymbol{E}}_1 \exp\{\mathrm{i}(2\boldsymbol{k}_0 - \boldsymbol{q})\cdot\boldsymbol{r} - \mathrm{i}(2\omega_0 - \delta_1)t\}$, oscillate at high frequency and do not survive the slow-timescale averaging. In eqn (6.42), we require $\nabla^2 |\bar{\boldsymbol{E}}|^2$. Although $|\bar{\boldsymbol{E}}_0(\boldsymbol{r}, t)|^2$ in eqn (6.87) has spatial dependence, it is relatively weak, and we may neglect $\nabla^2 |\bar{\boldsymbol{E}}_0|^2$ compared to $q^2 |\bar{\boldsymbol{E}}_{1,2}|^2$. Thus, to leading order, eqn (6.87) leads to

$$\nabla^2 |\bar{\boldsymbol{E}}|^2 = -\frac{q^2}{4}(\bar{\boldsymbol{E}}_0 \cdot \bar{\boldsymbol{E}}_1^* \mathrm{e}^{-\mathrm{i}\delta_1 t} + \bar{\boldsymbol{E}}_0^* \cdot \bar{\boldsymbol{E}}_2 \mathrm{e}^{\mathrm{i}\delta_2 t})\mathrm{e}^{\mathrm{i}\boldsymbol{q}\cdot\boldsymbol{r}} + \text{c.c.} \quad (6.88)$$

On the left-hand side of eqn (6.42), using eqn (6.70), we have

$$\left(\frac{\partial^2}{\partial t^2} - c_\mathrm{s}^2\nabla^2\right)\delta n = \frac{1}{2}\left(\frac{\partial^2}{\partial t^2} + c_\mathrm{s}^2 q^2\right) n_q \mathrm{e}^{\mathrm{i}\boldsymbol{q}\cdot\boldsymbol{r}} + \text{c.c.} \quad (6.89)$$

Substituting eqns (6.88) and (6.89) into eqn (6.42), and equating coefficients of $\exp(\mathrm{i}\boldsymbol{q}\cdot\boldsymbol{r})$, we obtain

$$\left(\frac{\partial^2}{\partial t^2} + c_\mathrm{s}^2 q^2\right) n_q = -\frac{\varepsilon_0 q^2}{8M}(\bar{\boldsymbol{E}}_0 \cdot \bar{\boldsymbol{E}}_1^* \mathrm{e}^{-\mathrm{i}\delta_1 t} + \bar{\boldsymbol{E}}_0^* \cdot \bar{\boldsymbol{E}}_2 \mathrm{e}^{\mathrm{i}\delta_2 t}). \quad (6.90)$$

This describes how the additional waves $\tilde{\boldsymbol{E}}_1$ and $\tilde{\boldsymbol{E}}_2$, generated by the non-linear coupling of $\tilde{\boldsymbol{E}}_0$ with δn, interact with $\tilde{\boldsymbol{E}}_0$ to affect the evolution of δn.

To close our system of equations, we note that the existence of $\tilde{\boldsymbol{E}}_1$ and $\tilde{\boldsymbol{E}}_2$ will produce terms in $\delta n\tilde{\boldsymbol{E}}$ additional to those considered at eqn (6.72). The contributions to $\delta n\tilde{\boldsymbol{E}}$ that vary as $\exp(\pm\mathrm{i}\boldsymbol{k}_0\cdot\boldsymbol{r})$ affect the evolution of $\tilde{\boldsymbol{E}}_0$ itself. Combining eqns (6.68), (6.70), (6.73), and (6.74) in eqn (6.21), and equating coefficients of $\exp(\mathrm{i}\boldsymbol{k}_0\cdot\boldsymbol{r})$, we obtain

$$\frac{1}{2}\left\{-\omega_0^2\bar{\boldsymbol{E}}_0 - 2\mathrm{i}\omega_0\frac{\partial\bar{\boldsymbol{E}}_0}{\partial t} + \omega_\mathrm{pe}^2\bar{\boldsymbol{E}}_0 + 3v_\mathrm{Te}^2 k_0^2\bar{\boldsymbol{E}}_0\right.$$
$$\left. + \frac{\omega_\mathrm{pe}^2}{2}\left(\frac{n_q}{n_0}\bar{\boldsymbol{E}}_1\mathrm{e}^{-\mathrm{i}(\omega_1-\omega_0)t} + \frac{n_q^*}{n_0}\bar{\boldsymbol{E}}_2\mathrm{e}^{-\mathrm{i}(\omega_2-\omega_0)t}\right)\right\} + \text{c.c.} = 0. \quad (6.91)$$

Using eqns (6.67), (6.81), (6.82), and (6.84), eqn (6.91) can be written

$$\mathrm{i}\frac{\partial\bar{\boldsymbol{E}}_0}{\partial t} = \frac{\omega_\mathrm{pe}}{4}\left(\frac{n_q}{n_0}\bar{\boldsymbol{E}}_1\mathrm{e}^{\mathrm{i}\delta_1 t} + \frac{n_q^*}{n_0}\bar{\boldsymbol{E}}_2\mathrm{e}^{\mathrm{i}\delta_2 t}\right). \quad (6.92)$$

The Zakharov equations have thus led to a closed set of equations, eqns (6.85), (6.86), (6.90), and (6.92), which together describe the coupled, non-linear evolution of the wave amplitudes $\bar{\boldsymbol{E}}_0$, n_q, $\bar{\boldsymbol{E}}_1$, and $\bar{\boldsymbol{E}}_2$. Comparing eqn (6.92) with eqns (6.85) and (6.86), and recalling $|\bar{\boldsymbol{E}}_0| \gg |\bar{\boldsymbol{E}}_{1,2}|$, we note that

$$\frac{1}{|\bar{\boldsymbol{E}}_{1,2}|}\left|\frac{\partial \bar{\boldsymbol{E}}_{1,2}}{\partial t}\right| \gg \frac{1}{|\bar{\boldsymbol{E}}_0|}\left|\frac{\partial \bar{\boldsymbol{E}}_0}{\partial t}\right|. \tag{6.93}$$

To good approximation, we may therefore treat $\bar{\boldsymbol{E}}_0$ in eqns (6.85) and (6.86) as a constant, while the other wave amplitudes evolve more rapidly. Furthermore, eqn (6.90) has the quasistatic solution

$$n_q = -\frac{\varepsilon_0}{8Mc_s^2}(\bar{\boldsymbol{E}}_0 \cdot \bar{\boldsymbol{E}}_1^* \mathrm{e}^{-\mathrm{i}\delta_1 t} + \bar{\boldsymbol{E}}_0^* \cdot \bar{\boldsymbol{E}}_2 \mathrm{e}^{\mathrm{i}\delta_2 t}). \tag{6.94}$$

This is the discrete-wave analogue of eqn (6.54). Substituting eqn (6.94) into eqns (6.85) and (6.86), and using eqn (6.83), we obtain

$$\mathrm{i}\frac{\partial \bar{\boldsymbol{E}}_1}{\partial t} = -\frac{\omega_{\mathrm{pe}}\varepsilon_0}{32Mn_0c_s^2}\bar{\boldsymbol{E}}_0(\bar{\boldsymbol{E}}_0^* \cdot \bar{\boldsymbol{E}}_1 + \bar{\boldsymbol{E}}_0 \cdot \bar{\boldsymbol{E}}_2^* \mathrm{e}^{2\mathrm{i}\mu t}), \tag{6.95}$$

$$\mathrm{i}\frac{\partial \bar{\boldsymbol{E}}_2}{\partial t} = -\frac{\omega_{\mathrm{pe}}\varepsilon_0}{32Mn_0c_s^2}\bar{\boldsymbol{E}}_0(\bar{\boldsymbol{E}}_0 \cdot \bar{\boldsymbol{E}}_1^* \mathrm{e}^{2\mathrm{i}\mu t} + \bar{\boldsymbol{E}}_0^* \cdot \bar{\boldsymbol{E}}_2). \tag{6.96}$$

Let us define the complex phase ϕ_0 of the approximately constant quantity $\bar{\boldsymbol{E}}_0$ by

$$\bar{\boldsymbol{E}}_0 = |\bar{\boldsymbol{E}}_0|\,\mathrm{e}^{\mathrm{i}\phi_0}. \tag{6.97}$$

It is also convenient to follow eqn (6.44), and define a turbulence parameter

$$\bar{W}_0 = \frac{\varepsilon_0\,|\bar{\boldsymbol{E}}_0|^2}{4n_0k_{\mathrm{B}}(T_{\mathrm{e}} + 3T_{\mathrm{i}})}. \tag{6.98}$$

We have included T_{i} in eqn (6.98) simply for convenience. Using eqns (6.41) and (6.98), the coupling coefficient in eqns (6.95) and (6.96) can be written

$$\frac{\omega_{\mathrm{pe}}\varepsilon_0}{32Mn_0c_s^2} = \frac{\omega_{\mathrm{pe}}\bar{W}_0}{8\,|\boldsymbol{E}_0|^2}. \tag{6.99}$$

For simplicity, we shall assume that the electric field amplitudes all lie along the same axis, so that we no longer need to write them as vector quantities. Then, using eqns (6.97) and (6.99), eqns (6.95) and (6.96) become

$$\mathrm{i}\frac{\partial \bar{E}_1}{\partial t} = -\frac{\omega_{\mathrm{pe}}\bar{W}_0}{8}(\bar{E}_1 + \bar{E}_2^* \mathrm{e}^{2\mathrm{i}(\phi_0+\mu t)}) \tag{6.100}$$

$$\mathrm{i}\frac{\partial \bar{E}_2}{\partial t} = -\frac{\omega_{\mathrm{pe}}\bar{W}_0}{8}(\bar{E}_1^* \mathrm{e}^{2\mathrm{i}(\phi_0+\mu t)} + \bar{E}_2). \tag{6.101}$$

It is convenient to define new variables

$$u_{1,2} = \bar{E}_{1,2} e^{-i(\phi_0 + \mu t)}, \tag{6.102}$$

so that

$$\frac{\partial \bar{E}_{1,2}}{\partial t} = \left(\frac{\partial u_{1,2}}{\partial t} + i\mu u_{1,2} \right) e^{i(\phi_0 + \mu t)}. \tag{6.103}$$

Multiplying eqns (6.100) and (6.101) by $\exp\{-i(\phi_0 + \mu t)\}$, and using eqns (6.102) and (6.103), we obtain

$$\frac{\partial u_1}{\partial t} + i\mu u_1 = i \frac{\omega_{pe} \bar{W}_0}{8} (u_1 + u_2^*) \tag{6.104}$$

$$\frac{\partial u_2}{\partial t} + i\mu u_2 = i \frac{\omega_{pe} \bar{W}_0}{8} (u_1^* + u_2). \tag{6.105}$$

This coupled linear system can be solved as an eigenvalue problem. We seek solutions of the form

$$u_1 = u_{10} e^{i\alpha t}, \qquad u_2^* = u_{20}^* e^{i\alpha t}, \tag{6.106}$$

where u_{10} and u_{20} are constant amplitudes. This yields

$$\alpha^2 = \mu \left(\mu - \frac{\omega_{pe} \bar{W}_0}{4} \right). \tag{6.107}$$

The condition for instability is that α^2 is negative, and hence α imaginary, so that by eqn (6.106) the time-dependence of u_1 and u_2 is exponential rather than oscillatory. From eqns (6.107) and (6.83), instability occurs if

$$\bar{W}_0 > \frac{6q^2}{k_D^2}. \tag{6.108}$$

This is the discrete-wave case of the modulational instability that we discussed in Section 6.2. Given a density perturbation of the form eqn (6.70), the electrostatic wave eqn (6.68) is unstable if its amplitude is sufficiently large that eqn (6.108) is satisfied, with $\bar{W}_0$ given by eqn (6.98). In this case, the non-linear coupling of the two waves, governed by the Zakharov equations, causes energy to flow from the initial $\tilde{\boldsymbol{E}}_0$ wave to $\tilde{\boldsymbol{E}}_1$ and $\tilde{\boldsymbol{E}}_2$, which are described by eqns (6.73) and (6.74). That is, the distribution of energy in wavevector space is altered. The simple fluid equations that we developed in Section 6.1 have enabled us to deal with the collective interaction of waves, which are themselves collective manifestations of plasma particle dynamics.

Exercises

6.1. The phenomenon of parametric resonance, which we examine in this exercise, provides a simple model for the interaction of two waves. Consider a one-dimensional harmonic oscillator whose natural frequency varies with time, such that the equation of motion is

$$\ddot{x}(t) + \omega_0^2[1 + \varepsilon \cos \alpha t]x(t) = 0.$$

There are two frequencies in this problem: ω_0, which is the frequency of the oscillator in the limit $\varepsilon \to 0$; and α, which is the frequency of the perturbation. The equation of motion may also be written

$$\ddot{x}(t) + \omega_0^2 x(t) = -\varepsilon\omega_0^2 x(t) \cos \alpha t,$$

where the right-hand side is effectively a wave-coupling term.

(a) For the case of approximate second harmonic resonance, we write $\alpha = 2\omega_0 + \eta$, where $\eta \ll \omega_0$. Bearing in mind the solution when $\varepsilon = 0$, we shall seek a solution of the form

$$x(t) = a(t) \cos[(\omega_0 + \eta/2)t] + b(t) \sin[(\omega_0 + \eta/2)t].$$

Obtain expressions for $\ddot{x}(t)$ and $x(t) \cos \alpha t$ in this case.

(b) Neglecting $\ddot{a}$, $\ddot{b}$ compared to $\omega_0\dot{a}$, $\omega_0\dot{b}$, and also the high-frequency terms introduced by wave interactions, show that

$$2\dot{b} - \eta a + \tfrac{1}{2}\varepsilon\omega_0 a = 0,$$

$$2\dot{a} + \eta b + \tfrac{1}{2}\varepsilon\omega_0 b = 0.$$

(c) Seek exponential solutions of the form $a(t), b(t) \sim \exp(\gamma t)$, and show that growth requires

$$-\tfrac{1}{2}\varepsilon\omega_0 < \eta < \tfrac{1}{2}\varepsilon\omega_0.$$

This relation between the amplitude ε and frequency $2\omega_0 + \eta$ of the perturbation gives the condition for parametric resonance instability; the resonant parameters are, of course, the two frequencies.

6.2. The nonlinear interaction of three waves can be represented by the following system of equations:

$$\frac{\partial u_1}{\partial t} - i\omega_1 u_1 = c_{23} u_2 u_3,$$

$$\frac{\partial u_2}{\partial t} - i\omega_2 u_2 = c_{13} u_1 u_3^*,$$

$$\frac{\partial u_3}{\partial t} - i\omega_3 u_3 = c_{12} u_1 u_2^*.$$

Here u_1, u_2, and u_3 are complex wave amplitudes, and c_{12}, c_{13}, and c_{23} are real coupling coefficients. By considering the small-amplitude limit, or the case where the coupling coefficients tend to zero, we see that ω_1, ω_2, and ω_3 are the real frequencies of linear normal modes.

(a) Express the system of equations in terms of the variables $v_j = u_j \exp(-i\omega_j t)$ for $j = 1, 2, 3$, using also the definition $\Omega = \omega_1 - \omega_2 - \omega_3$.

(b) Defining $a_j = |v_j|$, so that $v_j = a_j \exp(\mathrm{i}\phi_j)$ for $j = 1, 2, 3$, obtain the real system of equations

$$\frac{\partial a_1}{\partial t} = c_{23} a_2 a_3 \cos \Phi,$$

$$\frac{\partial a_2}{\partial t} = c_{13} a_1 a_3 \cos \Phi,$$

$$\frac{\partial a_3}{\partial t} = c_{12} a_1 a_2 \cos \Phi,$$

$$\frac{\partial \Phi}{\partial t} = \Omega - \left\{ c_{12} \frac{a_1 a_2}{a_3} + c_{13} \frac{a_1 a_3}{a_2} + c_{23} \frac{a_2 a_3}{a_1} \right\} \sin \Phi,$$

where

$$\Phi = \Omega t + \phi_2 + \phi_3 - \phi_1.$$

(c) Show that the following quantities are conserved:

$$M_1 = \frac{a_1^2}{c_{23}} - \frac{a_2^2}{c_{13}},$$

$$M_2 = \frac{a_2^2}{c_{13}} - \frac{a_3^2}{c_{12}}.$$

These are known as Manley–Rowe relations.

6.3. (a) Write down the first and second Zakharov equations for the case where only one spatial variable x is considered.

(b) Show that

$$\int_{-\infty}^{\infty} |\bar{\boldsymbol{E}}|^2 \, \mathrm{d}x$$

is a conserved quantity for the Zakharov equations.

(c) Demonstrate the equivalent relation for the case of three discrete electrostatic waves coupled by a density perturbation, namely that

$$|\bar{\boldsymbol{E}}_0|^2 + |\bar{\boldsymbol{E}}_1|^2 + |\bar{\boldsymbol{E}}_2|^2$$

is a conserved quantity for the Zakharov equations.

Solutions are on pages 156 *to* 158.

Solutions to exercises

1.1.

ω_{pe}	λ_D	N_D
5.7×10^{10}	6.9×10^{-6}	330
1.8×10^{8}	2.2×10^{-2}	4.4×10^{8}
5.7×10^{6}	2.2×10^{-2}	4.4×10^{5}
1.8×10^{11}	2.2×10^{-4}	4.4×10^{8}

1.2. Both ions and electrons will collide with the surface of the spacecraft, giving rise to electric currents whose densities are $J_i = n_{i0} e \bar{v}_i$ and $J_e = -n_{e0} e \bar{v}_e$ respectively, where n_{i0} and n_{e0} are the equilibrium number densities. Since the plasma is electrically neutral, $n_{i0} = n_{e0}$ and therefore $|J_e|/J_i = \bar{v}_e/\bar{v}_i \gg 1$. The flux of electrons greatly exceeds that of ions, so that the spacecraft will charge up rapidly. This will give the spacecraft an electrical potential V, which will be negative. As the magnitude of V increases, it will reduce the electron current, which will eventually fall to a level at which it is exactly cancelled by the ion current, so that no further charging occurs.

To estimate the maximum magnitude of V, we assume that the response of the surrounding plasma is dominated by the electrons, whose local number density becomes $n_e = n_{e0} \exp(eV/k_B T_e)$. The vanishing of the total current implies that

$$\bar{v}_e n_{e0} \exp(eV/k_B T_e) = \bar{v}_i n_{i0},$$

and the required expression follows immediately.

1.3. (a) The gravitational restoring force on the pendulum can be approximated by $-Mgx_n/l$, since $x_n/l \ll 1$. There are two springs attached to the bob, which exert forces proportional to their extension or compression. Thus

$$M\frac{d^2x_n}{dt^2} = -\frac{Mgx_n}{l} + K(x_{n+1} - x_n) - K(x_n - x_{n-1}).$$

(b) In the continuous approximation, $x_n(t)$ does not differ greatly from $x_{n+1}(t)$. It follows that $x(z, t)$ is not a rapidly varying function of z, so that if we Taylor expand the new expressions, truncation after a few terms should give an adequate description. Thus

$$x_{n-1}(t) \simeq x(z, t) - a\frac{\partial}{\partial z}x(z, t) + \frac{a^2}{2}\frac{\partial^2 x}{\partial z^2}(z, t),$$

$$x_{n+1}(t) \simeq x(z, t) + a\frac{\partial}{\partial z}x(z, t) + \frac{a^2}{2}\frac{\partial^2 x}{\partial z^2}(z, t).$$

Substituting these expansions into the result of (a), we obtain

$$\frac{\partial^2 x}{\partial t^2} = -\frac{g}{l}x + \frac{Ka^2}{M}\frac{\partial^2 x}{\partial z^2}.$$

This is known as the Klein–Gordon equation. If $x(z, t)$ is proportional to $\exp(2\pi iz/\lambda - i\omega t)$, eqn (1.20) follows.

(c) Returning to the wave equation that was derived in (b), let us write $x(z, t) = X(z)\cos\omega t$, where $\omega^2 < g/l$. Then we have

$$\frac{Ka^2}{M}\frac{d^2X}{dz^2} = (g/l - \omega^2)X.$$

Defining $\alpha = \{M(g/l - \omega^2)/Ka^2\}^{\frac{1}{2}}$, the general solution is

$$X(z) = A\exp(-\alpha z) + B\exp(\alpha z).$$

Let us choose our coordinates so that the first bob is at $z = 0$, and the line of bobs extends along the positive z-axis. Clearly $B = 0$, otherwise the value of X becomes unreasonably large when z is large. The constant A is equal to the amplitude of oscillation of the first bob, and

$$x(z, t) = A\exp(-\alpha z)\cos\omega t.$$

The amplitude of the displacement declines as we move along the line of bobs, becoming negligible where z exceeds a few times $1/\alpha$.

2.1. (a) 1.1×10^7 rad s^{-1}; (b) 54 rad s^{-1}; (c) 4.5×10^{10} rad s^{-1}

2.2. At $x = 0$, $y = 0$, $z = (2\pi p/qB)\cos\theta$. The parallel velocity $(p/m)\cos\theta$ is unaffected by the magnetic field. We know that the perpendicular motion is periodic, with angular frequency qB/m. It follows that after a time $2\pi m/qB$, the particle returns to $x = 0$ and $y = 0$, having travelled a distance $(2\pi m/qB)(p/m)\cos\theta$ along the z-axis.

2.3. Denoting velocities before and after the increase in magnetic field strength by subscripts 1 and 2 respectively, we have $v_{\perp 2}^2 = \alpha v_{\perp 1}^2$ by the conservation of magnetic moment, whereas $v_{\parallel 2}^2 = v_{\parallel 1}^2$. For a given electron, the initial and final energies are related by

$$\begin{aligned} K_2 &= (m/2)(v_{\parallel 2}^2 + v_{\perp 2}^2) = (m/2)(v_{\parallel 1}^2 + \alpha v_{\perp 1}^2) \\ &= K_1 + (m/2)(\alpha - 1)v_{\perp 1}^2. \end{aligned}$$

It is convenient to write $v_{\perp 1} = v_0 \sin\theta$, and since $K_1 = mv_0^2/2$, we have a change in energy

$$\Delta K = K_2 - K_1 = K_1(\alpha - 1)\sin^2\theta$$

for the electron considered. Initially, the electrons are uniformly distributed over an infinitesimally thin spherical shell in velocity space, with radius v_0. The number of electrons per unit area of this shell is $N/4\pi v_0^2$, and the number lying between θ and $\theta + \mathrm{d}\theta$ is accordingly

$$\frac{N}{4\pi v_0^2} 2\pi v_0^2 \sin\theta \,\mathrm{d}\theta = (N/2)\sin\theta\,\mathrm{d}\theta.$$

As a check, this gives the total number of electrons as

$$\int_0^{\pi} (N/2)\sin\theta\,\mathrm{d}\theta = N,$$

as required. Each electron initially at θ undergoes a change in energy $\Delta K(\theta)$, given by the formula above. Summing the change in energy of each electron, we obtain the total change in energy of the system:

$$\begin{aligned} \int_0^{\pi} \Delta K(\theta)(N/2)\sin\theta\,\mathrm{d}\theta &= \frac{NK_1(\alpha - 1)}{2}\int_0^{\pi} \sin^3\theta\,\mathrm{d}\theta \\ &= (2NK_1/3)(\alpha - 1), \end{aligned}$$

where NK_1 is the initial energy of the system.

3.1. (a) By eqn (3.42), the vector amplitude of the right circularly polarized wave can be written $E_{\mathrm{R}}(\hat{\boldsymbol{e}}_x + \mathrm{i}\hat{\boldsymbol{e}}_y)$—note that $|\hat{\boldsymbol{e}}_x + \mathrm{i}\hat{\boldsymbol{e}}_y| = \sqrt{2}$. Similarly, by eqn (3.39), the vector amplitude of the left circularly polarized wave can be written $E_{\mathrm{L}}(\hat{\boldsymbol{e}}_x - \mathrm{i}\hat{\boldsymbol{e}}_y)$. These waves oscillate as $\exp(\mathrm{i}k_{\mathrm{R}}z - \mathrm{i}\omega t)$ and $\exp(\mathrm{i}k_{\mathrm{L}}z - \mathrm{i}\omega t)$, respectively; summing them to form the resultant field, we obtain the required expression.

(b) From the expression derived in (a), it follows that

$$\frac{E_x}{E_y} = -\mathrm{i}\,\frac{\{E_{\mathrm{R}}\exp(\mathrm{i}k_{\mathrm{R}}z) + E_{\mathrm{L}}\exp(\mathrm{i}k_{\mathrm{L}}z)\}}{\{E_{\mathrm{R}}\exp(\mathrm{i}k_{\mathrm{R}}z) - E_{\mathrm{L}}\exp(\mathrm{i}k_{\mathrm{L}}z)\}}.$$

The required expression follows when we divide top and bottom by $E_{\mathrm{R}}\exp(\mathrm{i}k_{\mathrm{R}}z)$.

(c) We have $E_L = E_R$; in this case, multiplying top and bottom of the expression derived in (b) by $\exp\{i(k_R - k_L)z/2\}$, we obtain

$$\frac{E_x}{E_y} = -i\frac{[\exp\{i(k_R - k_L)z/2\} + \exp\{-i(k_R - k_L)z/2\}]}{[\exp\{i(k_R - k_L)z/2\} - \exp\{-i(k_R - k_L)z/2\}]}$$
$$= \cot\{(k_L - k_R)z/2\}.$$

This equation describes Faraday rotation. If E_x/E_y is infinite at $z = 0$, it has fallen to zero once the wave has travelled a distance L such that

$$(k_L - k_R)L/2 = \pi/2.$$

(d) Since $\omega_{pe}^2/\omega^2 \ll 1$, we may use the binomial theorem to give good approximations of the square roots in eqns (3.38) and (3.41). Thus

$$k_L \simeq \frac{\omega}{c}\left\{1 - \frac{\omega_{pe}^2}{2\omega^2(1 + \omega_{ce}/\omega)}\right\},$$

$$k_R \simeq \frac{\omega}{c}\left\{1 - \frac{\omega_{pe}^2}{2\omega^2(1 - \omega_{ce}/\omega)}\right\}.$$

Following (c), the angle of Faraday rotation is proportional to

$$k_L - k_R \simeq \frac{\omega_{ce}}{c}\frac{\omega_{pe}^2}{\omega^2 - \omega_{ce}^2}$$
$$\simeq \frac{\omega_{ce}}{c}\frac{\omega_{pe}^2}{\omega^2},$$

since $\omega^2 \gg \omega_{ce}^2$. By eqns (2.4) and (1.6), ω_{ce} is proportional to magnetic field strength and ω_{pe}^2 to electron density.

3.2. (a) By eqn (3.49),

$$\frac{\partial\omega}{\partial k} = \frac{2kc^2}{\omega_{pe}^2}\omega_{ce}.$$

(b) By eqn (3.49), $k = (\omega/\omega_{ce})^{1/2}\omega_{pe}/c$. Substituting this expression into the formula for the group velocity,

$$\frac{\partial\omega}{\partial k} = \frac{2c}{\omega_{pe}}(\omega\omega_{ce})^{1/2}.$$

Thus the group velocity is greater for higher-frequency Whistler waves, which accordingly arrive earlier.

4.1. (a) We start from Ohm's law eqn (4.10), which we use to replace $\boldsymbol{J}$ in eqn (4.26):

$$\nabla \times \boldsymbol{B} = \mu_0\sigma(\boldsymbol{E} + \boldsymbol{v} \times \boldsymbol{B}).$$

Taking the curl of this equation, and dividing by $\mu_0\sigma$, we obtain

$$\frac{1}{\mu_0\sigma}\nabla\times(\nabla\times\boldsymbol{B})=\nabla\times\boldsymbol{E}+\nabla\times(\boldsymbol{v}\times\boldsymbol{B}).$$

The left-hand side is dealt with using the general identity $\nabla\times(\nabla\times\boldsymbol{A}) = \nabla(\nabla\cdot\boldsymbol{A})-\nabla^2\boldsymbol{A}$, together with eqn (I.7); and the first term on the right-hand side using eqn (I.4). It follows that

$$\frac{\partial\boldsymbol{B}}{\partial t}=\nabla\times(\boldsymbol{v}\times\boldsymbol{B})+\frac{1}{\mu_0\sigma}\nabla^2\boldsymbol{B}.$$

Note that in the limit of infinite conductivity, this equation reduces to eqn (4.13), and we recover magnetic flux freezing.

(b) If the ideal term on the right-hand side of the previous equation is negligible, we are left with a diffusion equation:

$$\frac{\partial\boldsymbol{B}}{\partial t}=\frac{1}{\mu_0\sigma}\nabla^2\boldsymbol{B}.$$

The fact that $\boldsymbol{B}$ is a vector is not important, as each of its components separately satisfies the diffusion equation. We shall assume that each component is separable in x and t: that is, it can be written in the form $X(x)\,\theta(t)$. Then we have

$$X(x)\,\theta'(t)=\frac{1}{\mu_0\sigma}X''(x)\theta(t),$$

where the prime denotes differentiation with respect to x or t as appropriate. Dividing both sides by $X(x)\theta(t)$,

$$\frac{\theta'(t)}{\theta(t)}=\frac{1}{\mu_0\sigma}\frac{X''(x)}{X(x)}.$$

The left- and right-hand sides of this equation are functions of different independent variables. Equality for all values of x and t is possible only if both sides of the equation are, in fact, constant. Let us write

$$\frac{\theta'(t)}{\theta(t)}=\frac{1}{T},$$

where T is a fixed quantity which necessarily has time as its dimension. Similarly, we write

$$\frac{X''(x)}{X(x)}=\frac{1}{L^2},$$

where L is a fixed quantity which necessarily has length as its dimension. Substituting back, we obtain the relation

$$T=\mu_0\sigma L^2.$$

(c) It follows from the previous equation that small values of T, corresponding to rapid decay, occur when the resistivity is high (small σ) and the magnetic field gradient is steep (small L). Since we associate steep magnetic field gradients with large currents, through eqn (4.26), it is natural to deduce that the decay of the magnetic field is associated with Joule heating.

We can go further. The rate of working of the $\boldsymbol{J} \times \boldsymbol{B}$ force on unit volume of the fluid is

$$(\boldsymbol{J} \times \boldsymbol{B}) \cdot \boldsymbol{v} = -\boldsymbol{J} \cdot (\boldsymbol{v} \times \boldsymbol{B}).$$

Using eqn (4.10), it follows that

$$(\boldsymbol{J} \times \boldsymbol{B}) \cdot \boldsymbol{v} = -\frac{J^2}{\sigma} + \boldsymbol{J} \cdot \boldsymbol{E}.$$

This is the non-ideal generalisation of eqn (4.97). The steps between eqns (4.99) and (4.104) apply also to the non-ideal case. Using eqn (4.104), it follows from the preceding equation that

$$(\boldsymbol{J} \times \boldsymbol{B}) \cdot \boldsymbol{v} + \frac{J^2}{\sigma} = -\boldsymbol{\nabla} \cdot \boldsymbol{S} - \frac{\partial}{\partial t}\left(\frac{B^2}{2\mu_0}\right).$$

Thus, the energy drawn from the magnetic field energy density and from the Poynting flux goes both to support the rate of working of the $\boldsymbol{J} \times \boldsymbol{B}$ force, as in the ideal case, and the non-ideal dissipative term J^2/σ. The latter can be identified with the Joule heating encountered in basic electrical theory, as follows. Recall that electrical resistance $R = \rho l/A$, where $\rho = 1/\sigma$ and l and A are length and area respectively. Then, since $J = I/A$ by definition, where I is the electrical current, we have

$$\frac{J^2}{\sigma} = \frac{I^2 R}{Al}.$$

We recognise I^2R as the rate of Joule heating in a resistor R carrying current I, so that J^2/σ is the rate of Joule heating per unit volume.

4.2. Following Section 4.2, we set out to balance the ram pressure of the solar wind against the pressure of the Earth's magnetic field. Our estimate of the position of the magnetopause is given by the value of R at which the balance occurs:

$$\tfrac{1}{2} n_{\mathrm{sw}} m_{\mathrm{p}} v_{\mathrm{sw}}^2 = \frac{B^2}{2\mu_0},$$

where m_p is the mass of the proton (1.67×10^{-27} kg) and $\mu_0 = 4\pi \times 10^{-7}\ \mathrm{Hm}^{-1}$. Substituting the values of n_{sw} and v_{sw} already given, together with the formula for B, we find that the distance to the magnetopause is approximately ten Earth radii.

4.3. (a) Clearly, the magnetic field is azimuthal: $\boldsymbol{B} = B(r)\hat{\boldsymbol{e}}_\theta$, where $\hat{\boldsymbol{e}}_\theta$ is the azimuthal unit vector. We shall use Ampère's law eqn (4.26), together with Stokes' theorem eqn (I.5). Integrating over a disc S of radius r, which lies in the plane perpendicular to the z-axis, we obtain

$$\int_S (\boldsymbol{\nabla} \times \boldsymbol{B}) \cdot \mathrm{d}\boldsymbol{S} = \oint \boldsymbol{B} \cdot \mathrm{d}\boldsymbol{l} = 2\pi r B(r)$$

$$= \mu_0 \int_S \boldsymbol{J} \cdot \mathrm{d}\boldsymbol{S} = \mu_0 \int_0^r J(r')\, 2\pi r'\, \mathrm{d}r'.$$

Hence

$$B(r) = \frac{\mu_0}{r} \int_0^r J(r') r'\, \mathrm{d}r'.$$

(b) Take the derivative of the preceding expression with respect to r:

$$\frac{\mathrm{d}B}{\mathrm{d}r} = -\frac{\mu_0}{r^2} \int_0^r J(r')\, r'\, \mathrm{d}r' + \mu_0 J(r).$$

Then

$$\mu_0 J(r) = \frac{B(r)}{r} + \frac{\mathrm{d}B}{\mathrm{d}r}.$$

(c) $\boldsymbol{J} \times \boldsymbol{B} = J(r)\hat{\boldsymbol{e}}_z \times B(r)\hat{\boldsymbol{e}}_\theta = -J(r)B(r)\hat{\boldsymbol{e}}_r$. This is an inward radial force, whose magnitude

$$J(r)B(r) = \frac{1}{\mu_0}\left(\frac{B^2}{r} + B\frac{\mathrm{d}B}{\mathrm{d}r}\right),$$

using the result of (b). It is easy to check that this expression is equivalent to the more compact one given in the question.

(d) At equilibrium, by eqn (4.6), $\boldsymbol{J} \times \boldsymbol{B} = \boldsymbol{\nabla} p$. That is, using the previous result,

$$\frac{\mathrm{d}p}{\mathrm{d}r} = \frac{-1}{2\mu_0 r^2} \frac{\mathrm{d}}{\mathrm{d}r} \{r^2 B^2(r)\}.$$

Integrating this expression outwards from $r' = r$ to $r' = a$, where the pressure is zero, we obtain

$$p(r) = \frac{1}{2\mu_0} \int_r^a \frac{1}{r'^2} \frac{\mathrm{d}}{\mathrm{d}r'} \{r'^2 B^2(r')\}\, \mathrm{d}r',$$

where $B(r)$ follows from (a) above.

5.1. Multiplying eqn (5.9) by the electron mass m, and integrating over all velocities, we obtain

$$\int m \frac{\partial f}{\partial t} \mathrm{d}^3\boldsymbol{v} + \int m\boldsymbol{v} \cdot \frac{\partial f}{\partial \boldsymbol{x}} \mathrm{d}^3\boldsymbol{v} + q \int (\boldsymbol{E} + \boldsymbol{v} \times \boldsymbol{B}) \cdot \frac{\partial f}{\partial \boldsymbol{v}} \mathrm{d}^3\boldsymbol{v} = 0.$$

The first integral is simply

$$\int m \frac{\partial f}{\partial t} \, \mathrm{d}^3 \boldsymbol{v} = \frac{\partial}{\partial t} \int mf \, \mathrm{d}^3 \boldsymbol{v} = \frac{\partial \rho}{\partial t}.$$

The second integral is

$$\int \frac{\partial}{\partial \boldsymbol{x}} \cdot (m\boldsymbol{v}f) \, \mathrm{d}^3 \boldsymbol{v} = \frac{\partial}{\partial \boldsymbol{x}} \cdot \int m\boldsymbol{v}f \, \mathrm{d}^3 \boldsymbol{v}$$

since, as noted after eqn (5.6), $\frac{\partial}{\partial \boldsymbol{x}} \cdot \boldsymbol{v}$ vanishes. Thus, we already have the two terms required in eqn (4.1). It remains to show that the third integral above vanishes. First,

$$\int \boldsymbol{E} \cdot \frac{\partial f}{\partial \boldsymbol{v}} \, \mathrm{d}^3 \boldsymbol{v} = \int \frac{\partial}{\partial \boldsymbol{v}} \cdot (f\boldsymbol{E}) \, \mathrm{d}^3 \boldsymbol{v} = \int f\boldsymbol{E} \cdot \mathrm{d}\boldsymbol{S}_{\mathrm{v}}.$$

Here we have used the fact that $\boldsymbol{E}$ is independent of $\boldsymbol{v}$, followed by the divergence theorem in velocity space—recall eqn (I.2). We have denoted the surface element in velocity space by $\mathrm{d}\boldsymbol{S}_{\mathrm{v}}$. Next, we employ a standard trick: because we have integrated over all velocities, the surface of integration lies at $|\boldsymbol{v}| \to \infty$, where f is infinitesimally small, so that the integral vanishes. Finally, by Leibniz' rule,

$$\int (\boldsymbol{v} \times \boldsymbol{B}) \cdot \frac{\partial f}{\partial \boldsymbol{v}} \, \mathrm{d}^3 \boldsymbol{v} = \int \frac{\partial}{\partial \boldsymbol{v}} \cdot \{(\boldsymbol{v} \times \boldsymbol{B}) f\} \, \mathrm{d}^3 \boldsymbol{v} - \int f \left\{ \frac{\partial}{\partial \boldsymbol{v}} \cdot (\boldsymbol{v} \times \boldsymbol{B}) \right\} \mathrm{d}^3 \boldsymbol{v}.$$

The first integral on the right-hand side can be shown to vanish by the arguments that we have just employed for the integral involving $\boldsymbol{E}$; the second integral vanishes because of the vector identity $(\partial / \partial \boldsymbol{v}) \cdot (\boldsymbol{v} \times \boldsymbol{B}) = 0$.

5.2. The normalization of f_0 is governed by eqn (5.16), which implies

$$\int_0^{v_{\mathrm{b}}} \frac{A}{v_{\mathrm{c}}^3 + v^3} \, \mathrm{d}^3 \boldsymbol{v} = n_0.$$

Since $f_0(v)$ is isotropic, we can write the volume element in velocity space simply as $\mathrm{d}^3 \boldsymbol{v} = 4\pi v^2 \, \mathrm{d}v$, so that

$$A \int_0^{v_{\mathrm{b}}} \frac{v^2 \, \mathrm{d}v}{v_{\mathrm{c}}^3 + v^3} = \frac{n_0}{4\pi}.$$

The integration can be carried out using the substitution $u = v_{\mathrm{c}}^3 + v^3$, since v_{c} is constant. This gives

$$\frac{A}{3} \int_{v_{\mathrm{c}}^3}^{v_{\mathrm{c}}^3 + v_{\mathrm{b}}^3} \frac{\mathrm{d}u}{u} = \frac{n_0}{4\pi},$$

and therefore

$$A = (3n_0/4\pi)\{\ln(1 + v_b^3/v_c^3)\}^{-1}.$$

5.3. (a) Starting from eqn (5.7), we note that since the gravitational acceleration $\dfrac{d\boldsymbol{v}}{dt} = -\nabla\Phi(\boldsymbol{x})$, we have $\dfrac{\partial}{\partial \boldsymbol{v}} \cdot \dfrac{d\boldsymbol{v}}{dt} = 0$, and therefore

$$\frac{\partial f}{\partial t} + \boldsymbol{v} \cdot \frac{\partial f}{\partial \boldsymbol{x}} - \frac{\partial \Phi}{\partial \boldsymbol{x}} \cdot \frac{\partial f}{\partial \boldsymbol{v}} = 0.$$

(b) Integrating the previous equation over all velocity space, the first two terms immediately give $\dfrac{\partial \nu}{\partial t}$ and $\dfrac{\partial}{\partial \boldsymbol{x}} \cdot (\nu \bar{\boldsymbol{v}})$, as in Exercise 5.1. The remaining term is

$$\int \frac{\partial \Phi}{\partial \boldsymbol{x}} \cdot \frac{\partial f}{\partial \boldsymbol{v}} \, d^3\boldsymbol{v} = \frac{\partial \Phi}{\partial \boldsymbol{x}} \cdot \int \frac{\partial f}{\partial \boldsymbol{v}} \, d^3\boldsymbol{v}.$$

On performing the velocity integration here, we obtain zero, since there are no stars with infinite velocities. Thus

$$\frac{\partial \nu}{\partial t} + \frac{\partial}{\partial \boldsymbol{x}} \cdot (\nu \bar{\boldsymbol{v}}) = 0.$$

(c) We multiply the galactic Vlasov equation by v_j, and integrate over all velocity space to obtain

$$\int v_j \frac{\partial f}{\partial t} \, d^3\boldsymbol{v} + \int v_j v_i \frac{\partial f}{\partial x_i} \, d^3\boldsymbol{v} - \int v_j \frac{\partial \Phi}{\partial x_i} \frac{\partial f}{\partial v_i} \, d^3\boldsymbol{v} = 0.$$

Using our earlier definitions, this can be written

$$\frac{\partial}{\partial t}(\nu \bar{v}_j) + \frac{\partial}{\partial x_i}(\nu \overline{v_i v_j}) - \frac{\partial \Phi}{\partial x_i} \int v_j \frac{\partial f}{\partial v_i} \, d^3\boldsymbol{v} = 0.$$

The first two terms are as required. In the final term, we note that $\dfrac{\partial v_j}{\partial v_i} = \delta_{ij}$, where the Kronecker delta δ_{ij} is defined after eqn (4.39). Then

$$v_j \frac{\partial f}{\partial v_i} = \frac{\partial}{\partial v_i}(v_j f) - \delta_{ij} f.$$

When we integrate over all velocities, the first term on the right-hand side vanishes as usual, and we are left with

$$\int v_j \frac{\partial f}{\partial v_i} \, d^3\boldsymbol{v} = -\delta_{ij} \int f \, d^3\boldsymbol{v} = -\delta_{ij} \nu.$$

It follows that

$$\frac{\partial \Phi}{\partial x_i} \int v_j \frac{\partial f}{\partial v_i} \, d^3\boldsymbol{v} = -\nu \frac{\partial \Phi}{\partial x_j},$$

as required.

6.1. (a) Using Leibniz' rule,

$$\ddot{x}(t) = \{\ddot{a} + 2(\omega_0 + \eta/2)\dot{b} - (\omega_0 + \eta/2)^2 a\} \cos[(\omega_0 + \eta/2)t] \\ + \{\ddot{b} - 2(\omega_0 + \eta/2)\dot{a} - (\omega_0 + \eta/2)^2 b\} \sin[(\omega_0 + \eta/2)t].$$

Using the standard trigonometric identities $\cos A \cos B = \frac{1}{2}[\cos(A + B) + \cos(A - B)]$ and $\cos A \sin B = \frac{1}{2}[\sin(A + B) - \sin(A - B)]$, we obtain

$$x(t) \cos \alpha t = \tfrac{1}{2}a(t)[\cos(3\alpha t/2) + \cos(\alpha t/2)] \\ + \tfrac{1}{2}b(t)[\sin(3\alpha t/2) - \sin(\alpha t/2)],$$

where $\alpha/2 = \omega_0 + \eta/2$.

(b) Using the approximations given, together with $\eta \ll \omega_0$, the results of (a) become

$$\ddot{x}(t) \simeq [2\dot{b} - (\omega_0 + \eta)a]\omega_0 \cos[(\omega_0 + \eta/2)t] \\ - [2\dot{a} + (\omega_0 + \eta)b]\omega_0 \sin[(\omega_0 + \eta/2)t],$$

$$x(t) \cos \alpha t \simeq \tfrac{1}{2}a \cos[(\omega_0 + \eta/2)t] - \tfrac{1}{2}b \sin[(\omega_0 + \eta/2)t].$$

Substituting these expressions into the original equation of motion gives

$$(2\dot{b} - \eta a + \tfrac{1}{2}\varepsilon\omega_0 a) \cos[(\omega_0 + \eta/2)t] \\ - (2\dot{a} + \eta b + \tfrac{1}{2}\varepsilon\omega_0 b) \sin[(\omega_0 + \eta/2)t] = 0.$$

The required expressions follow, since sine and cosine are independent functions whose coefficients must vanish separately.

(c) We have

$$\begin{bmatrix} 2\gamma & \tfrac{1}{2}\varepsilon\omega_0 + \eta \\ \tfrac{1}{2}\varepsilon\omega_0 - \eta & 2\gamma \end{bmatrix} \begin{bmatrix} a \\ b \end{bmatrix} = \begin{bmatrix} 0 \\ 0 \end{bmatrix}.$$

The determinant must vanish, so that

$$4\gamma^2 = (\tfrac{1}{2}\varepsilon\omega_0)^2 - \eta^2.$$

Instability requires γ positive, and hence $\gamma^2 > 0$, from which the required condition follows.

6.2. (a) Substitution gives

$$\frac{\partial v_1}{\partial t} = c_{23} v_2 v_3 \exp(\mathrm{i}\Omega t),$$

$$\frac{\partial v_2}{\partial t} = c_{13} v_1 v_3^* \exp(-\mathrm{i}\Omega t),$$

$$\frac{\partial v_3}{\partial t} = c_{12} v_1 v_2^* \exp(-\mathrm{i}\Omega t).$$

(b) It follows from the equations obtained in (a) that

$$\frac{\partial a_1}{\partial t} + \mathrm{i}a_1 \frac{\partial \phi_1}{\partial t} = c_{23}a_2a_3 \exp(\mathrm{i}\Phi),$$

$$\frac{\partial a_2}{\partial t} + \mathrm{i}a_2 \frac{\partial \phi_2}{\partial t} = c_{13}a_1a_3 \exp(-\mathrm{i}\Phi),$$

$$\frac{\partial a_3}{\partial t} + \mathrm{i}a_3 \frac{\partial \phi_3}{\partial t} = c_{12}a_1a_2 \exp(-\mathrm{i}\Phi).$$

The three evolution equations for the a_j follow from the real parts of the above expressions. The evolution equation for Φ follows when we combine the imaginary parts of the above expressions.

(c) It follows from the expressions obtained in (b) that

$$\frac{a_1}{c_{23}} \frac{\partial a_1}{\partial t} = \frac{a_2}{c_{13}} \frac{\partial a_2}{\partial t} = \frac{a_3}{c_{12}} \frac{\partial a_3}{\partial t} = a_1a_2a_3 \cos\Phi;$$

then, by the definitions of M_1 and M_2,

$$\frac{\partial M_1}{\partial t} = \frac{\partial M_2}{\partial t} = 0.$$

6.3. (a) From eqns (6.28) and (6.42),

$$\mathrm{i}\omega_{\mathrm{pe}} \frac{\partial \bar{\boldsymbol{E}}}{\partial t} + \frac{3}{2} v_{\mathrm{Te}}^2 \frac{\partial^2 \bar{\boldsymbol{E}}}{\partial x^2} = \frac{\omega_{\mathrm{pe}}^2}{2} \frac{\delta n}{n_0} \bar{\boldsymbol{E}},$$

$$\frac{\partial^2}{\partial t^2} \frac{\delta n}{n_0} - c_{\mathrm{s}}^2 \frac{\partial^2}{\partial x^2} \frac{\delta n}{n_0} = \frac{\varepsilon_0}{4Mn_0} \frac{\partial^2}{\partial x^2} |\bar{\boldsymbol{E}}|^2.$$

(b) We wish to show that

$$\frac{\mathrm{d}}{\mathrm{d}t} \int_{-\infty}^{\infty} |\bar{\boldsymbol{E}}|^2 \,\mathrm{d}x \equiv \int_{-\infty}^{\infty} \frac{\partial}{\partial t} |\bar{\boldsymbol{E}}|^2 \,\mathrm{d}x = 0.$$

Since $|\bar{\boldsymbol{E}}|^2 = \bar{\boldsymbol{E}} \cdot \bar{\boldsymbol{E}}^*$, we may use Leibniz' rule to write

$$\frac{\partial}{\partial t} |\bar{\boldsymbol{E}}|^2 = \bar{\boldsymbol{E}} \cdot \frac{\partial \bar{\boldsymbol{E}}^*}{\partial t} + \bar{\boldsymbol{E}}^* \cdot \frac{\partial \bar{\boldsymbol{E}}}{\partial t}.$$

Using the first equation in (a), and noting that δn is a real quantity, it follows that

$$\frac{\partial}{\partial t} |\bar{\boldsymbol{E}}|^2 = -\frac{\mathrm{i}}{\omega_{\mathrm{pe}}} \left\{ \bar{\boldsymbol{E}}^* \cdot \left(\frac{\omega_{\mathrm{pe}}^2}{2} \frac{\delta n}{n_0} \bar{\boldsymbol{E}} - \frac{3}{2} v_{\mathrm{Te}}^2 \frac{\partial^2 \bar{\boldsymbol{E}}}{\partial x^2} \right) \right.$$
$$\left. - \bar{\boldsymbol{E}} \cdot \left(\frac{\omega_{\mathrm{pe}}^2}{2} \frac{\delta n}{n_0} \bar{\boldsymbol{E}}^* - \frac{3}{2} v_{\mathrm{Te}}^2 \frac{\partial^2 \bar{\boldsymbol{E}}^*}{\partial x^2} \right) \right\} =$$

$$= \frac{3\mathrm{i}}{2}\frac{v_{\mathrm{Te}}^2}{\omega_{\mathrm{pe}}}\left(\bar{\boldsymbol{E}}^* \cdot \frac{\partial^2 \bar{\boldsymbol{E}}}{\partial x^2} - \bar{\boldsymbol{E}} \cdot \frac{\partial^2 \bar{\boldsymbol{E}}^*}{\partial x^2}\right)$$

$$= \frac{3\mathrm{i}}{2}\frac{v_{\mathrm{Te}}^2}{\omega_{\mathrm{pe}}}\frac{\partial}{\partial x}\left(\bar{\boldsymbol{E}}^* \cdot \frac{\partial \bar{\boldsymbol{E}}}{\partial x} - \bar{\boldsymbol{E}} \cdot \frac{\partial \bar{\boldsymbol{E}}^*}{\partial x}\right).$$

If we now operate on both sides of this equation with $\int_{-\infty}^{\infty} \mathrm{d}x$, the right-hand side vanishes for the usual reason: the fields and their gradients vanish at $\pm\infty$.

(c) Using eqns. (6.85), (6.86), and (6.92), we obtain

$$\frac{\partial}{\partial t}|\bar{\boldsymbol{E}}_1|^2 = \bar{\boldsymbol{E}}_1^* \cdot \frac{\partial \bar{\boldsymbol{E}}_1}{\partial t} + \bar{\boldsymbol{E}}_1 \cdot \frac{\partial \bar{\boldsymbol{E}}_1^*}{\partial t}$$

$$= -\frac{\mathrm{i}\omega_{\mathrm{pe}}}{4n_0}(n_q^* \bar{\boldsymbol{E}}_1^* \cdot \bar{\boldsymbol{E}}_0 \mathrm{e}^{-\mathrm{i}\delta_1 t} - n_q \bar{\boldsymbol{E}}_1 \cdot \bar{\boldsymbol{E}}_0^* \mathrm{e}^{\mathrm{i}\delta_1 t}),$$

$$\frac{\partial}{\partial t}|\bar{\boldsymbol{E}}_2|^2 = -\frac{\mathrm{i}\omega_{\mathrm{pe}}}{4n_0}(n_q \bar{\boldsymbol{E}}_2^* \cdot \bar{\boldsymbol{E}}_0 \mathrm{e}^{-\mathrm{i}\delta_2 t} - n_q^* \bar{\boldsymbol{E}}_2 \cdot \bar{\boldsymbol{E}}_0^* \mathrm{e}^{\mathrm{i}\delta_2 t}),$$

$$\frac{\partial}{\partial t}|\bar{\boldsymbol{E}}_0|^2 = -\frac{\mathrm{i}\omega_{\mathrm{pe}}}{4n_0}(n_q \bar{\boldsymbol{E}}_0^* \cdot \bar{\boldsymbol{E}}_1 \mathrm{e}^{\mathrm{i}\delta_1 t} + n_q^* \bar{\boldsymbol{E}}_0^* \cdot \bar{\boldsymbol{E}}_2 \mathrm{e}^{\mathrm{i}\delta_2 t}$$

$$- n_q^* \bar{\boldsymbol{E}}_0 \cdot \bar{\boldsymbol{E}}_1^* \mathrm{e}^{-\mathrm{i}\delta_1 t} - n_q \bar{\boldsymbol{E}}_0 \cdot \bar{\boldsymbol{E}}_2^* \mathrm{e}^{-\mathrm{i}\delta_2 t}).$$

Summing these expressions, we recover the required result.

Index

Bold page numbers denote a definition